The Patrick Moore Practical

More information about this series at http://www.springer.com/series/3192

Your Guide to the 2017 Total Solar Eclipse

Michael E. Bakich

 Springer

Michael E. Bakich
Astronomy Magazine
Milwaukee, WI, USA

ISSN 1431-9756   ISSN 2197-6562   (electronic)
The Patrick Moore Practical Astronomy Series
ISBN 978-3-319-27630-4   ISBN 978-3-319-27632-8   (eBook)
DOI 10.1007/978-3-319-27632-8

Library of Congress Control Number: 2016936671

This Springer imprint is published by Springer Nature
The registered company is Springer International Publishing AG Switzerland

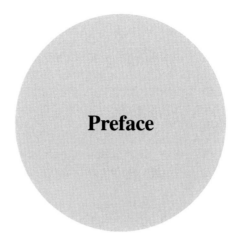

Preface

The original meaning of ἔκλειψη (the Greek word for "eclipse") is a forsaking, quitting, or disappearance. Hence the covering over of one object by another or the immersion of something into something else represents precisely the facts of an eclipse.

Earth and the Moon are solid bodies in space. Each casts a shadow as a result of the Sun's illumination. To understand eclipses, all we need to know is what results from the existence of these shadows. Total eclipses, be they of the Sun or the Moon, are examples of sublime celestial geometry. Each one is an exact lineup of the Sun, the Moon, and Earth for a total solar eclipse, or the Sun, Earth, and the Moon for a total lunar eclipse.

Our solar system is a group of a huge number of bodies, a few large and many small. The main one is the Sun. Its Latin name, *sol*, indicates why we call the collection a solar system. Now imagine a line between the Sun and any other body at a given time. Because everything in the solar system is in motion, that line will point in a different direction as time passes. Such a line shows the direction of the object's shadow, precisely opposite the Sun's position in space.

Every so often, an additional body comes into alignment with the other two. If the two non-solar bodies are close enough, the shadow from the closest one to the Sun may fall on the other. It may completely cover the second body or only partially cover it. Likewise, the first body may completely block out the Sun's disk or it may only partially obscure it. It is during these times that eclipses occur.

The larger a body is, the farther into space it will cast its shadow. At Earth's average distance from the Sun, any object casts an umbral shadow 108 times its diameter. This makes Earth's umbral shadow an average length of 855,000 miles and the Moon's umbral shadow approximately 255,000 miles long. Of course, these numbers vary because the distances of these bodies from the Sun change. Still, with them in mind it's easy to see why total lunar eclipses last much longer than total solar eclipses. The disk of Earth's shadow is much larger than the corresponding disk of the Moon's shadow at the average Earth-Moon distance of about 238,900 miles.

Most readers of this book will never have experienced a total solar eclipse and may therefore think that solar eclipses are rare. Actually, at least two and as many as five occur every year. During the period from 2000 B.C. to 3000 A.D., a total of 11,898 solar eclipses occurred. Of that number 3,173 (26.7 percent) were total. Within that span of five millennia, Earth experienced five solar eclipses in one calendar year only 25 times (0.5 percent). The most recent was in 1935, and the next time will not be until 2206.

Indeed the numbers surrounding eclipses, the scientific reasons they happen, and the way astronomers can predict—to a fraction of a second—where, when, and for how long a given eclipse will occur make these events fascinating. But all of this pales in comparison to actually witnessing totality at your location.

94.5 percent of the continental United States will experience a partial eclipse on August 21, 2017. Do you know the difference between a partial eclipse and a total one? It's the difference between a lightning bug and lightning. Between testing negative and positive with a pregnancy test. Between a paper cut and stepping on a land mine. In other words, there's no comparison.

Thankfully, comets and eclipses no longer generate the anxiety and alarm among uneducated populations that they did even as recently as a century ago. This means the upcoming total solar eclipse on August 21, 2017, will not only attract a good deal of attention from many millions of people, it may even induce a respectable number to think about the science and history of eclipses. And that's a good thing. Every now and then when something this remarkable happens—a great thunderstorm, an earthquake, a volcanic eruption, a bright comet, or an eclipse—it allows people who normally don't think of astronomy a chance to stop and appreciate the wonderful universe we live in.

Because this book is about a scientific event, it contains lots of facts. But it's also meant to appeal to astronomy newbies, people who certainly will be interested in this event, but who may not be well versed in science.

That said, my advice regarding how you use this book is to concentrate on the section or sections that mean the most to you when you're ready to deal with them. Your first order of business probably will be either to familiarize yourself with what's actually going to happen or to identify the ideal location you'd like to be at on eclipse day. Later, you may be interested in reading about eye safety. At some point, you'll want to cross-reference your ideal location with the discussion of weather prospects you'll read about here. And if you wish to enhance your viewing, the equipment chapters will speak to you as the event approaches.

However you choose to approach it, I cannot stress enough that you really should observe the eclipse. This is a must-see event. I think of it as "awesome" in the truest sense of that word: able to inspire or generate awe. Especially in the United States, people throw that word around like it's nothing "Your shoes are awesome!" "This crème brûlée is awesome!" "Little Julie's new crayon drawing is awesome!" Really? Do these things actually generate *awe*? On second thought, probably not.

But the eclipse on August 21, 2017, will be nothing short of awe-inspiring. I guarantee that if you stand under the Moon's shadow in the daytime you'll never forget it. Furthermore, it will stand out as one of the greatest—if not the greatest—sights you ever have or ever will behold. I'm smiling as I write this because I know some of you are thinking, "Wow, this guy should have worked for P. T. Barnum!"

Remember, however, that I've traveled to observe 13 total solar eclipses, and for 11 of those, I had groups accompanying me. I made passionate presentations to thousands of people before those events. And afterward, how many people thought I'd gone overboard? That I'd over-hyped the eclipse? That I'd set their expectations so high they could never reach them?

Zero.

Milwaukee, WI, USA   Michael E. Bakich

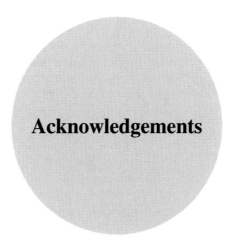

Acknowledgements

First, as always, I want to thank my lovely wife, Holley, for both her emotional and real support with regard to this book. The latter manifests itself in the many illustrations she created, and the former because of her wonderful attitude of, "Sure, I can do that."

I also am indebted to David J. Eicher, editor of *Astronomy* magazine, for letting me cherry-pick all the "non-Holley" eclipse-related illustrations our magazine has produced over the years. This gold mine of gorgeous explanatory material goes back more than four decades.

As far as the actual photography of eclipses, I called on two of the best. Astronomy professor and *Astronomy* magazine Contributing Editor Mike Reynolds (coauthor of my most recent book) has seen 18 total solar eclipses and photographed them all. And I think he's sent every image he ever took to me to use as I please. Thanks, pal.

Ben Cooper, who maintains the excellent photographic website LaunchPhotography.com, started photographing total solar eclipses on August 1, 2008, during a trip he accompanied me on to Novosibirsk, Russia. His shots were great then and, to my surprise, they've actually gotten better. You rock, dude!

Thanks also to eclipse meteorologist Jay Anderson. His willingness to share his weather predictions for the 2017 eclipse made Chapter 24 far more accurate than it would have been. The entire cadre of eclipse chasers worldwide owes Jay a huge debt. Keep up the great work, sir!

I want to thank Kate Russo for sharing her "Community Eclipse Planning" white paper that I turned into Chapter 23. Want to talk eclipses? She is a wealth of information who offers her services as an eclipse planner. That's right, she not only talks the talk, she walks the walk, too.

Finally, I want to thank everyone at Springer involved with this project. To Maury Solomon, who replied to my initial enquiry in less than a day, thanks for realizing this event was going to be so huge and that we needed to get this project going quickly. And if anyone rates an even bigger thanks, it's my point-of-contact editor, Nora Rawn. She answered every one of my questions immediately, thereby saving me any wasted effort. Nora also performed a meticulous edit on the manuscript. That said, I accept all responsibility for any factual errors. This book contains lots of numbers. Let's hope I got them all correct.

Contents

Chapter 1

What's All the Fuss About?

Drama is coming to the United States. But it won't be in the form of an economic collapse, a papal visit, or a political upheaval. On August 21, 2017, Sun-watchers along a thin curved line that stretches for thousands of miles from Oregon to South Carolina will experience nature's grandest spectacle: a total solar eclipse.

It's not a stretch to say that this might prove to be the most viewed sky event in history. That's why prior to the eclipse, astronomy clubs, government agencies, cities—even whole states—are preparing for the unprecedented onslaught of visitors seeking to experience darkness at midday.

© Springer International Publishing Switzerland 2016
M.E. Bakich, *Your Guide to the 2017 Total Solar Eclipse*, The Patrick Moore
Practical Astronomy Series, DOI 10.1007/978-3-319-27632-8_1

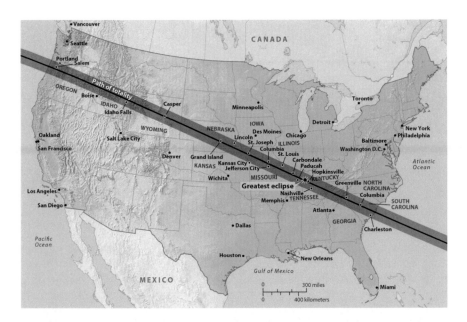

Fig. 1.1 The path of totality across the U.S. (Courtesy of *Astronomy* magazine: Richard Talcott and Roen Kelly)

One thing you may hear when the media starts to hype this eclipse is the name they ascribe to it. You'll hear it called "America's eclipse," "the United States' eclipse," "our eclipse," and similar monikers. Why? Does the U.S. have some special connection to this eclipse? Believe it or not, the answer is yes. The Moon's dark inner shadow, which is the only place the eclipse will be total, touches no other land on Earth outside the United States for this eclipse. It speeds along thousands and thousands of miles of open ocean waters—in the Pacific before it contacts land in Oregon, and then in the Atlantic after leaving land in South Carolina—but if you want to stand on solid ground and look up at the wonder in the sky, you'll need to be in America.

This will be the first total solar eclipse crossing the continental U.S. in 38 years. One did cover Hawaii in 1991, but in the 48 states the last one occurred February 26, 1979. Unfortunately, not many people saw it because it was visible from just five states in the Northwest. Making matters worse, that winter's weather for the most part was bleak along the path of totality. Before the 1979 eclipse, you have to go back to March 7, 1970, when a total solar eclipse traveled up the East Coast of the United States, again occurring in a scant five states. More people certainly saw that one—I'm one of them!—but because it happened 45 years ago, the percentage of people that you'll encounter who saw it will be essentially zero.

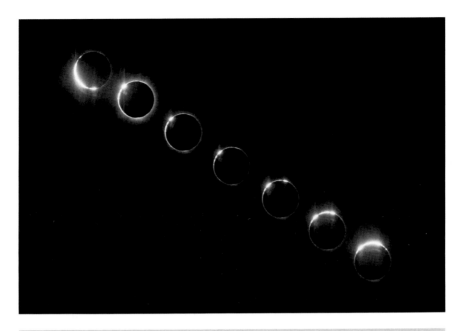

Fig. 1.2 This spectacular photographic sequence around totality shows some of the features you'll see August 21, 2017. (Courtesy of Ben Cooper)

Although total solar eclipses occur more often than total lunar ones, more people—actually, pretty much everyone—have seen a total eclipse of the Moon. Few, on the other hand, have seen a total solar eclipse. The reason is quite simple. We live on Earth, and it's our perspective that interacts with the geometry of these events. During a lunar eclipse, anyone on the night side of our planet under a clear sky can see the Moon passing through Earth's dark inner shadow. That shadow, even as far away as the Moon, is quite a bit larger than the Moon, so it takes our satellite some time to pass through it. In fact, if the Moon passes through the center of Earth's shadow the total part of the eclipse can last as much as 106 minutes. Usually totality doesn't last that long because the Moon passes either slightly above or below the center of the shadow our planet casts.

Conversely, the Moon and its shadow at the distance of Earth are much smaller; so small, in fact, that the shadow barely reaches our planet's surface. Anybody in the lighter outer region of the shadow will see a partial solar eclipse. The lucky individuals under the dark inner shadow will *experience*—a much better word than "see"—a total solar eclipse. But not for long. Solar totality lasts a maximum of 7½ minutes. In fact, the longest totality in the 5,000-year span from 2000 B.C. to 3000 A.D. is 7 minutes and 29 seconds. That eclipse will occur July 16, 2186.

Fig. 1.3 The Moon's penumbra (*lighter outer circle*) has an average diameter of approximately 4,350 miles (7,000 kilometers) at Earth's distance The umbra (here represented by the *small yellow circle* to make its size easier to see) has a maximum diameter of 166 miles. (Earth image courtesy of NASA; graphics: Holley Y. Bakich)

We won't be nearly that fortunate August 21, 2017. The maximum duration of totality then will be 2 minutes and 40 seconds. Ah, but what a span it will be!

Now, I do want to say a few words about the importance of totality. ***It's all about totality.*** Everyone in the continental U.S. will see at least a partial eclipse. In fact, if you have clear skies on eclipse day, the Moon's shadow will cover at least 48 percent of the Sun's surface. And that's from the northern tip of Maine. Although the Moon covering part of the Sun's disk sounds cool, you need to aim higher.

Likening a partial eclipse to a total eclipse is like comparing almost dying to dying. If you are outside during a solar eclipse with 48 percent coverage, you won't even notice your surroundings getting dark. And it doesn't matter whether the partial eclipse above your location is 48, 58, or 98 percent. Only totality reveals the true celestial spectacles: the two diamond rings, the Sun's glorious corona, 360° of sunset colors, and stars in the daytime. So, remember, to see any of this you must be in the path of totality.

Knowing that, you want to try to be on the centerline. The fact that the Moon's shadow is round probably isn't a revelation. If it were square, it wouldn't matter where you viewed totality. People across its width would experience the same duration of darkness. The shadow is round, however, so the longest eclipse occurs at its centerline because that's where you'll experience the Moon's shadow's full width.

Oh, there's something else. This event will happen! As astronomers (professional or amateur), some of the problems we deal with are due to the uncertainty and limited visibility of some celestial events. Comets may appear bright if their compositions are just so. Meteor showers might reach storm levels if we pass through a thick part of the stream (and generally the best views occur after midnight). A supernova as bright as a whole galaxy may be visible, but you need a telescope to view it. In contrast to such events, this solar eclipse will occur at the exact time astronomers predict, along a precisely plotted path, and for the lengths of time given. Guaranteed. Oh, and it's a daytime event to boot.

The next total solar eclipse that causes the Moon's shadow to fall across the continental U.S. occurs April 8, 2024. It's going to be a good one, too. Depending on where you are on the centerline, the duration of totality will last at least 3 minutes and 22 seconds on the east coast of Maine and stretches to 4 minutes and 27 seconds in southwestern Texas. After that eclipse, it's a 20-year wait until August 23, 2044 (and, similar to the 1979 event, that one is visible only in Montana and North Dakota). Total solar eclipses follow in 2045 and 2078. You can read about these upcoming events in Appendix A.

But it's 2017 that's causing all the excitement now. And you've got the right reference for it in your hands. I'll discuss trip planning, how to observe the event, top locations for activities and viewing, and much more. I'll weave interviews in, too, with people who are experts on photographing eclipses and others who will be conducting major observing events across the country. This book will keep you informed so that you can approach the eclipse without a shadow of doubt.

Chapter 2

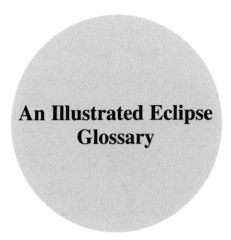

An Illustrated Eclipse Glossary

So that we're all speaking the same language about solar eclipses, this chapter will provide a brief list of the most popular terms you'll encounter, many of them with illustrations. You should get familiar with them because you will see them again.

Altitude—the height, in degrees, of a point or celestial object above the horizon. We measure altitude from 0° (on the horizon) to 90° (at the zenith, which is the overhead point). Consider the following sentence as an eclipse-related example: At Rosecrans Memorial Airport in St. Joseph, Missouri, on August 21, 2017, the Sun will stand 61.9° high in the south at mid-eclipse.

© Springer International Publishing Switzerland 2016
M.E. Bakich, *Your Guide to the 2017 Total Solar Eclipse*, The Patrick Moore
Practical Astronomy Series, DOI 10.1007/978-3-319-27632-8_2

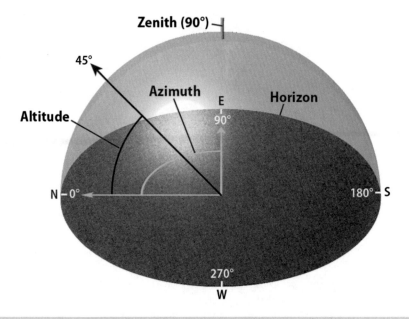

Fig. 2.1 Altitude and azimuth. (Courtesy of *Astronomy* magazine: Roen Kelly)

Angular diameter—the apparent size of a celestial object, measured in degrees, minutes, and/or seconds, as seen from Earth. OK, let's define the three words in that sentence. A degree is 1/360 of a circle. Said another way, a circle contains 360°. A minute (short for minute of arc or arcminute) is 1/60 of 1°. A second (short for second of arc or arcsecond) is 1/60 of 1 minute of arc. So, 1° contains 3,600 arcseconds. An example of use might be something like, "The average angular size of the Sun or the Moon, as seen from Earth, is 31 arcminutes, or 0.52°."

Angular distance—this is the same thing as angular diameter except that we're measuring the distance between two objects, not the size of a single object; so the definition would be the distance between two celestial bodies expressed in degrees, minutes, and/or seconds of arc.

Aphelion—the position of an object in solar orbit when it lies farthest from the Sun. Similarly, **apogee** is the position of the Moon or other object in Earth orbit when it lies farthest from our planet. Aphelion has two approved pronunciations: a FEEL ee on and ap HEEL ee on.

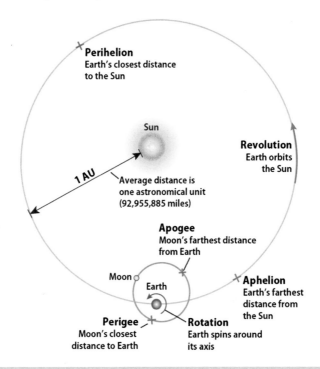

Fig. 2.2 Aphelion and perihelion. (Courtesy of *Astronomy* magazine: Richard Talcott and Roen Kelly)

Azimuth—the angular distance (from 0° to 360°) to an object measured eastward along the horizon starting from north; so the azimuth of an object due north is 0°; due east would be 90°; south would be 180°; and a due-west azimuth equals 270°. Well, that's if the object is on the ground; for a celestial object, the measurement is to a line that passes through the object and makes a right angle to the horizon.

Baily's beads—during a total solar eclipse, the effect often seen just before and just after totality when only a few points of sunlight are visible at the edge of the Moon. This effect is caused by the irregularity of the lunar surface. At our satellite's edge, mountains block out the Sun's disk, but valleys permit it to shine through. Scientists named this phenomenon after English astronomer Francis Baily, who first explained it in 1836.

Center line—the midpoint of the width of the Moon's shadow on Earth; the centerline is the location for the maximum duration of totality. You'll hear the cry throughout this book: "Get to the center line!"

Chromosphere—the region of the Sun's atmosphere between its visible surface and its corona; sometimes briefly visible just before or after totality as an intense red glow at the Moon's edge.

Conjunction—a point on the sky where two celestial bodies appear to line up; the lineup may be an exact one, as in the case of a total eclipse, or it may be a near one, as in the case of New Moon (when our satellite is "in line" with the Sun).

Corona—the shell of thin gas that extends out some distance from the Sun's surface normally visible only during totality; "corona" is the Latin word for "crown." Well put.

Fig. 2.3 The Sun's corona becomes visible during totality. (Courtesy of Mike Reynolds)

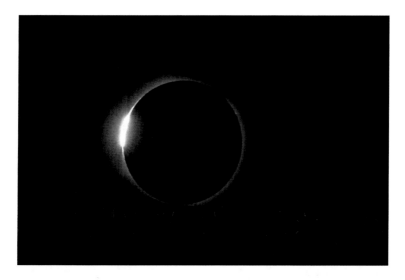

Fig. 2.4 The diamond ring is visible just before and just after totality. (Courtesy of Mike Reynolds)

Diamond ring—the effect just prior to or just after totality of a solar eclipse when a small portion of the Sun's disk plus its corona produce an effect similar to a ring with a brilliant diamond.

Disk—the visible surface of any heavenly body.

Ecliptic—the circle described by the Sun's apparent annual path through the stars; the plane of Earth's orbit around the Sun. You may not know it by this name, but the ecliptic traces the Sun's path through the constellations of the zodiac.

First contact—during a solar eclipse, the moment that the Moon makes contact with the Sun; this moment marks the beginning of the eclipse.

Flare—a sudden burst of particles and energy from the Sun's photosphere; through a Hydrogen-alpha filter, flares often appear brighter than the surrounding area.

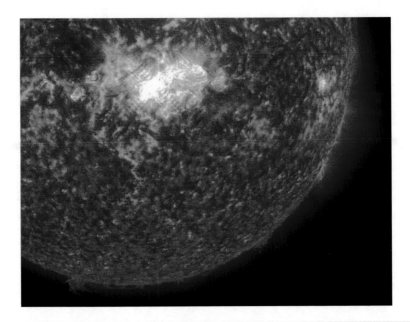

Fig.2.5 Solar flares are huge explosions on the Sun's surface. (Courtesy of NASA/SDO)

Fourth contact—during a solar eclipse, the moment that the disk of the Moon breaks contact with the Sun; this moment marks the end of the eclipse.

Hydrogen-alpha filter—a filter that passes only light with a wavelength of 656.28 nanometers (or 6,562.8 Angstroms); a simpler definition is a filter that allows you to observe the Sun's chromosphere, flares, prominences, and more; abbreviated H-alpha filters, these accessories are expensive but impressive.

Magnitude—the amount of the Sun's diameter the Moon covers during an eclipse; this is not the same as "obscuration."

New Moon—the phase where the Moon seems completely unlit from our perspective on Earth; the phase where the Moon is between Earth and the Sun; solar eclipses can occur only at New Moon.

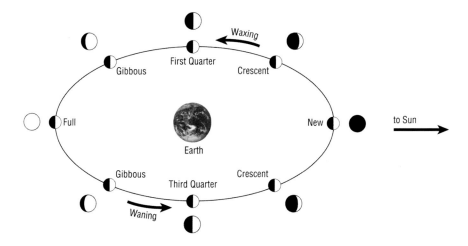

Fig. 2.6 Lunar phases. Solar eclipses occur only at New Moon. (Courtesy of Holley Y. Bakich)

Fig. 2.7 The Moon's umbra is its dark inner shadow; its penumbra is the lighter outer shadow. (Courtesy of Holley Y. Bakich)

Nodes—with regard to solar eclipses, the two points at which the Moon's orbital plane intersects the plane of the ecliptic; in other words, the two places the plane of the Moon's orbit crosses the plane of Earth's orbit; eclipses can occur only near nodes.

Obscuration—the amount of the Sun's area the Moon covers during an eclipse; this is not the same as "magnitude."

Orbit—the path of one celestial body around another; examples: Earth orbits the Sun, and the Moon orbits Earth.

Penumbra—the less dark outer region of the Moon's shadow; an observer under the penumbra sees a partial solar eclipse.

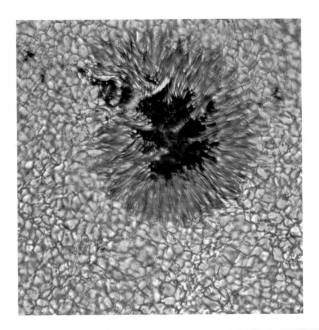

Fig. 2.8 Close-up view of a sunspot. (Courtesy of Vacuum Tower Telescope/NSO/NOAO)

Perigee—the position of the Moon or other object in Earth orbit when it lies closest to our planet.

Perihelion—the position of an object in solar orbit when it lies closest to the Sun.

Photosphere—the visible surface of the Sun; where our star emits visible light; the Sun's disk.

Prominence—a large-scale, gaseous formation above the surface of the Sun usually occurring over regions of solar activity such as sunspot groups; during totality observers often see prominences seeming to erupt from the Moon's dark edge.

Revolution—in astronomy, the orbiting of one body around another; the Moon revolves around Earth.

Rotation—the spinning of a celestial body on its axis; Earth rotates once a day.

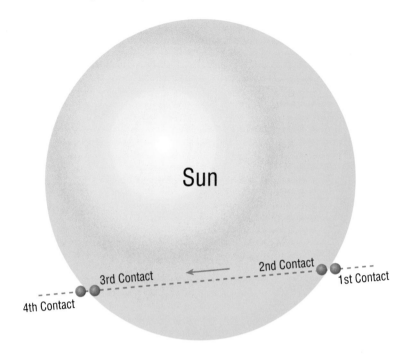

Fig. 2.9 First, second, third, and fourth contacts during a total solar eclipse have the same geometries as the planet pictured crossing the Sun in this illustration. (Courtesy of *Astronomy* magazine: Roen Kelly)

Saros cycle—a time period equal to 6,585.3 days between which similar eclipses occur.

Second contact—during a total solar eclipse, the moment the Moon covers 100 percent of the Sun's disk; the instant totality begins.

Shadow bands—faint ripples of light occasionally seen on flat, light–colored surfaces just before and just after totality.

Solar telescope—a telescope whose design lets you safely observe the Sun.

Sunspot—a temporarily cooler (and therefore darker) region on the Sun's visible disk caused by magnetic field variations.

Syzygy—the lineup of three celestial bodies; for a solar eclipse, the lineup is the Sun, the Moon, and Earth.

Third contact—during a total solar eclipse, the instant totality ends.

Umbra—the dark inner region of the Moon's shadow; anyone under the Moon's umbra will experience a total solar eclipse.

Universal Time (UT)—also known as Greenwich Mean Time (GMT); standard time kept on the Greenwich meridian (longitude=0°); astronomers use UT to coordinate observations of celestial events.

Chapter 3

Frequently Asked Questions Answered About Eclipses

The purpose of this chapter is to answer some of the most important questions for both the general public and the media. Yes, the eclipse is a year away, but it's never too early for knowledge, right? Plus, these are the facts, and they won't change.

As described earlier, a solar eclipse is a lineup of the Sun, the Moon, and Earth—in that order. The Moon, directly between the Sun and Earth, casts a shadow on our planet. If you're in the dark part of that shadow, called the umbra, you'll see a total eclipse. If you're in the light part, the penumbra, you'll see a partial eclipse.

Now the most-popular follow-up question: since the Sun larger than the Moon, how does this work? While our daytime star's diameter is approximately 400 times larger than that of the Moon, it also lies roughly 400 times farther away. This means both disks appear to our eyes to be the same size.

The next question is a frequent one: When do solar eclipses occur? A solar eclipse happens at New Moon. The Moon has to be between the Sun and Earth for a solar eclipse to occur. The only lunar phase when that happens is New Moon.

So then, why don't solar eclipses happen at every New Moon? The reason is that the Moon's orbit tilts 5° to Earth's orbit around the Sun. Astronomers call the two intersections of these paths nodes. Eclipses only occur when the Sun lies at one node and the Moon is at its New phase for solar eclipses or Full phase for lunar eclipses. During most lunar months, the Sun lies either above or below one of the nodes, and no eclipse happens.

© Springer International Publishing Switzerland 2016
M.E. Bakich, *Your Guide to the 2017 Total Solar Eclipse*, The Patrick Moore
Practical Astronomy Series, DOI 10.1007/978-3-319-27632-8_3

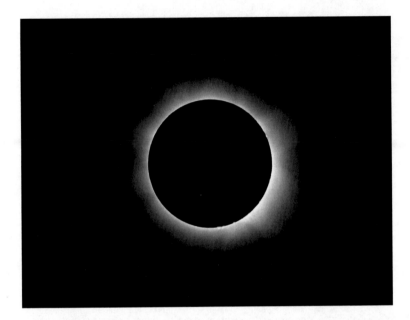

Fig. 3.1 Stand under the Moon's umbra during a total solar eclipse, and you'll experience the awe of totality. (Courtesy of Mike Reynolds)

Fig. 3.2 Although solar eclipses occur at New Moon, our satellite must lie at one of its nodes for the lineup to produce the event. (Courtesy of Holley Y. Bakich)

Another question people ask a lot is "Why are some eclipses longer than others?" The reason the total phases of solar eclipses vary in time is because Earth is not always at the same distance from the Sun, and the Moon is not always the same distance from Earth. The Earth-Sun distance varies by as much as 3 percent. That may not sound like much, but it's nearly 3 million miles. The Moon-Earth distance, meanwhile, can change by as much as 12 percent. The result is that while the Moon retains the same size at all times, the Moon's apparent diameter—that is, the disk that we see—can range from 7 percent larger than the Sun to 10 percent smaller than the Sun.

Next we have one about wording: What do magnitude and obscuration mean? Astronomers categorize each solar eclipse in terms of two properties, its magnitude and the percentage of obscuration, and I don't want you to be confused when you encounter these terms. The magnitude of a solar eclipse is the percent of the Sun's diameter that the Moon covers during maximum eclipse. The obscuration is the percent of the Sun's total surface area covered at maximum.

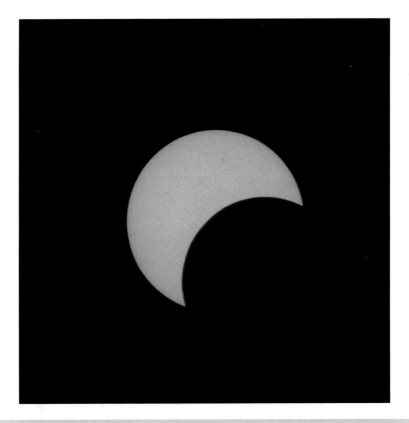

Fig. 3.3 This image shows the Moon covering half the Sun's diameter. The magnitude of the eclipse at this time would be 0.50, or 50 percent. The obscuration, however, that is, the percentage of the Sun's area covered, would be only 39.1 percent. (Courtesy of Ben Cooper)

Here's an example: Let's say we just observed a partial eclipse where the Moon covered half the Sun's diameter. In this case, the magnitude of the eclipse equals 50 percent. However, the amount of obscuration (that is, the area of the Sun's disk that the Moon blotted out) was only 39.1 percent. Now listen to this next part carefully. For a total solar eclipse, the obscuration (that is, the area of the Sun covered) always equals exactly 100 percent. You really can't cover more than 100 percent of the Sun's surface. The magnitude, however, can be anywhere from 100 percent, which astronomers would designate as 1.0000, to a bit more than 108 percent, or 1.0805, the magnitude of the total solar eclipse July 16, 2186. Magnitudes greater than 100 percent simply mean that the Moon's apparent diameter is that much greater than the Sun's. For the August 21, 2017 eclipse, the magnitude will be 1.0306.

Next up, what does the word "Saros" mean? This answer will be a bit technical because I want you to be familiar with it in case you encounter the term between now and the eclipse (I will discuss it further in Chapter 5 as well). This is the length of time between similar solar and lunar eclipses and represents their periodicity and recurrence, which repeat every 6,585.3 days. That length of time can be 14 regular years plus 4 leap years plus 11 days and 8 hours, or 13 regular years plus 5 leap years plus 10 days and 8 hours. Two eclipses separated by one Saros cycle are similar. They occur with the Sun, Earth, and Moon at the same relative positions. Also, the Moon's distance from Earth is nearly the same, and the eclipses happen at the same time of year. British astronomer Edmond Halley coined the term "saros" from an 11th-century Byzantine book.

The next question is what's the path of the eclipse through the U.S.? Here are just a few facts. The first place the Moon's shadow makes with land is on the Pacific Coast at Government Point, Oregon. After that, the centerline crosses through 12 states: Oregon, Idaho, Wyoming, Nebraska, Kansas, Missouri, Illinois, Kentucky, Tennessee, North Carolina, Georgia, and South Carolina. But the eclipse actually will be total in parts of 14 states. Tiny—and I mean really small—portions of Montana and Iowa do experience short spans of totality, but both regions are at the extremes of the umbra's path, not near the center line. The last point of contact in the U.S. is at the Atlantic Ocean's edge just southeast of Key Bay, South Carolina.

How long will totality be? On August 21, 2017, totality lasts a maximum of 2 minutes and 40.2 seconds. That's it. To experience that length, you'll need to be slightly south of Carbondale, Illinois, in Giant City State Park. You might think about getting there early. Actually, you might think about not going there at all because it probably will be swarming with people and non-moving vehicles. There are plenty of locations that offer more space, better service facilities, and a totality within a second or two of maximum. Rosecrans Airport in St. Joseph, Missouri, is one such a place.

The next question is on most people's list: besides totality, what else should observers look for? Although the big payoff is the exact lineup of the Sun, the Moon, and your location, keep your eyes open during the partial phases that lead up to and follow it. Cool things are afoot before and after totality. As you view the beginning of the eclipse through a safe solar filter, around the three-quarters mark, you'll start to notice that your shadow, and those of others, is getting sharper. The reason is that the Sun's disk is shrinking, literally approaching a point, and a smaller light source produces better-defined shadows. At about 85 percent coverage, Venus will be visible 34° west-northwest of the Sun. If any trees are at your viewing site, you may see their leaves act like pinhole cameras as hundreds of crescent Suns appear in their shadows.

Fig. 3.4 Around the time the Moon covers three-quarters of the Sun's disk, notice how your shadows (and those of everything around you) are starting to sharpen. (Courtesy of Holley Y. Bakich)

Fig. 3.5 Although you won't need any equipment to view totality, I highly recommend you bring binoculars, which will give you an unforgettable view. (Courtesy of Holley Y. Bakich)

Here's a question nobody can answer now, except in general terms: How many people will see this eclipse? My opinion is that this eclipse will be the most-viewed ever, and probably the number one most-watched sky event in history. Four factors will make it so: (1) the attention it will get from the media; (2) the superb coverage of the highway system in our country; (3) the typical weather throughout the country on August 21; and (4) the vast number of people who will have access to it from nearby large cities.

Safety questions are also sure to arise. What eye protection is needed during totality? In fact, totality is safe to look at. During the time the Moon's disk covers that of the Sun, it's perfectly fine to look at the eclipse. In fact, to experience the awesomeness of the event, you *must* look at the Sun without a filter during totality.

And the follow-up question, is any equipment required? A telescope is *not* needed. One of the great things about the total phase of a solar eclipse is that it looks best to naked eyes. The sight of the corona surrounding the Moon's black disk in a darkened sky is unforgettable. That said, binoculars give you a close-up view— but still at relatively low power—that you should take certainly advantage of during the event.

And our final question, what else will be happening as a result of the eclipse? Nature will take heed. Depending on your surroundings, as totality nears you may experience strange things. Look: you'll notice a resemblance to the onset of night, though not exactly. Areas much lighter than the sky near the Sun lie all around the horizon. Shadows also look different. Listen: usually, any breeze will dissipate and birds (many of whom will come in to roost) will stop chirping. It is quiet. Feel: a 10°–15° Fahrenheit drop in temperature is not unusual.

Chapter 4

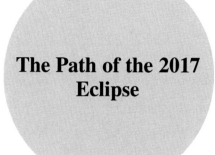

The Path of the 2017 Eclipse

This chapter provides you with a general overview of the path of the Moon's shadow across the U.S. during the eclipse. While eclipse predictions are exact, I didn't want to go crazy with decimal points or fractions. So, I have rounded off the times to the nearest second. Trust me, it won't make a bit of difference.

The first question to answer is, "Where in the U.S. does totality begin?" The Moon's umbra, after sweeping eastward across part of the Pacific Ocean, makes its initial landfall in Oregon. If you want to be the first person to experience totality on the centerline in the continental U.S., then be on the waterfront at Pacific Palisades, Oregon, at 10:15:57 A.M. PDT. There, the total phase lasts 1 minute, 59 seconds. The umbra remains in the state a bit more than 11 minutes, until 10:27:07 A.M. PDT. If you happen to be in Salem (the state capital), totality will begin at 10:17:21 and last for 1 minute 54 seconds.

What's the path after Oregon? The centerline crosses through 11 more states. After a great west-to-east path through its first state, the centerline takes roughly 12 minutes to cross a wide swath of Idaho, entering the western part of the state just before 11:25 A.M. MDT and leaving just before 11:37 A.M. MDT. The shadow hits only a few small cities in the eastern part of the state. Idaho Falls gets 1 minute and 50 seconds of totality. If, however, you drive north on U.S. 20 to the centerline just south of Rexburg, your time under the umbra will increase to 2 minutes and 18 seconds. An extra 28 seconds may not sound like it's worth the effort. Trust me. It is. No matter how long totality lasts for you during the 2017 eclipse, it won't be long enough.

© Springer International Publishing Switzerland 2016
M.E. Bakich, *Your Guide to the 2017 Total Solar Eclipse*, The Patrick Moore Practical Astronomy Series, DOI 10.1007/978-3-319-27632-8_4

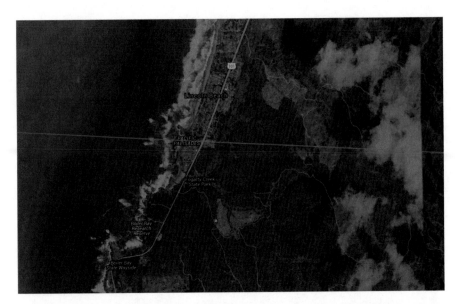

Fig. 4.1 The center line of the August 21, 2017, total solar eclipse first makes landfall at Pacific Palisades, Oregon. (Map courtesy of Xavier M. Jubier; data courtesy of Google Imagery/ DigitalGlobe/Landsat/State of Oregon)

Next up is Wyoming, where the umbral centerline dwells until just past 11:49 A.M. MDT. The big question for this state is, "How many people will fit into Grand Teton National Park?" Well, at least the southern part because that's where the eclipse's centerline passes through it. With the Sun's altitude riding at 50° during totality, you can bet many enterprising photographers will capture our eclipsed star above some great earthly terrain throughout the event. Locals might head to Riverton because the Moon's shadow passes between it and Boysen State Park. Take State Route 789 north out of town, and find a nice place off the road. There you'll enjoy 2 minutes and 24 seconds of totality.

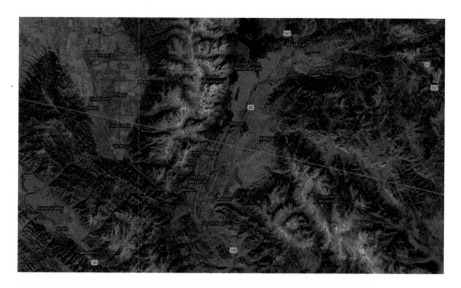

Fig. 4.2 The umbra passes through Grand Teton National Park. Several tour providers already have made Jackson, Wyoming, their destination on eclipse day. (Map courtesy of Xavier M. Jubier; data courtesy of Google Imagery/TerraMetrics)

The second actual city (after Salem, Oregon) to be rocked by totality is Casper, Wyoming's capital. People headed there—and believe me, lots of people are headed there—will experience 2 minutes and 27 seconds of totality. And don't despair if you're stuck in the city's outskirts. Totality at those locations lasts just 1 second less than the maximum in Casper. Not into Casper? The next locations east of there with good viewing prospects are Glendo, Glendo State Park, and Glendo Reservoir. Head there and you'll probably find fewer people enjoying a totality 2½ minutes long. Just be sure not to position yourself too close to a tree line that might block your view of any part of the eclipse, not just totality. You may not have heard of these locations, but they're easy to reach because they lie right along Interstate 25.

At 11:46:55 A.M. MDT, the eclipse begins along the centerline in the great state of Nebraska. Now listen to this next part carefully. When the Moon's shadow enters Nebraska, it's Mountain Daylight Time. One-third of the way through the state, however, it crosses into Central Daylight Time. Please, please, please remember this if you're headed to Nebraska. I'm not worried about residents because they're used to dealing with the time zones. Visitors enamored with the celestial lineup, however, may be another matter entirely. Totality along the centerline ends in Nebraska at 1:07:19 P.M. CDT.

Fig. 4.3 The United States Penitentiary in Leavenworth, Kansas, is among the more unusual locations the Moon's inner shadow will cover. (Courtesy of Americasroof/Wikimedia Commons)

The centerline hits the very northeastern part of Kansas at 1:04 P.M. CDT and enters Missouri a scant two minutes later. That makes it sound like nothing's happening in the Sunflower State, but actually the reverse is true. Small towns near the centerline will enjoy nearly the longest possible duration of totality. How does 2 minutes and 38 seconds sound for Hiawatha? The tiny burgs of Troy, Wathena, and Elwood will get an extra second over that. Atchison isn't on the centerline, but it still enjoys 2 minutes and 18 seconds under the Moon's umbra. Oh, and there's another town not on the centerline, Leavenworth (the site of the United States Penitentiary), which gets 1 minute and 41 seconds of darkness.

Finally, at 1:06:16 P.M. CDT, the Moon's dark inner shadow arrives at eclipse central: Missouri. Because of the irregularities in state boundaries, it re-enters Kansas four seconds later. Six seconds after that, it's back in Missouri.

My guess is that more people will see the eclipse in this state than any other. In fact, because two huge metropolitan areas—Kansas City and St. Louis—lie along the path's limits, millions of people will experience darkness in the daytime just by default. About half of each of those cities lies in the path (perhaps a bit less for St. Louis), but no part of either is on the centerline. Therefore readers should make the short drive north of Kansas City or south of St. Louis to head for a location near the centerline.

Lots of towns lie in the umbra's path through Missouri as well as two other sizeable cities: St. Joseph and Columbia, the home of the University of Missouri Tigers.

My company, Front Page Science, will be conducting a huge event at Rosecrans Memorial Airport in St. Joseph. We expect 100,000 people to join us there (yes, we have room), and perhaps thousands of others at secondary venues. Head to www.stjosepheclipse.com for more info about that specific event.

Fig. 4.4 The author will be conducting a huge, free public event at Rosecrans Memorial Airport in St. Joseph, Missouri. (Courtesy of the author)

If you want to be by yourself or with a small group for the eclipse instead of within a large group, head south out of St. Joseph along the 20-mile stretch of U.S. Route 169 to Gower. You'll experience 2 minutes and 39 seconds of totality all the way. Plattsburg, Lathrop, and Lawson all enjoy the same duration of darkness. If you head east from St. Joseph along U.S. Route 36, you'll have a bit more than 2½ minutes of totality all the way to the intersection of State Route 33. You can continue along U.S. 36 to Cameron, which lies a bit north of the centerline. Still this town's no slouch with 2 minutes and 24 seconds of totality.

Progressing along the centerline's path through Missouri, Carrollton will see 2 minutes and 36 seconds of totality and Marshall and Boonville, which sit a bit closer to mid-path, each get 2 minutes and 40 seconds. And then it's Columbia, MO's turn. While the city is not directly on the centerline, students, instructors, and others at the University of Missouri can gaze safely at the totally eclipsed Sun for 2 minutes and 36 seconds. For three additional seconds, head south a bit to Rock Bridge Memorial State Park. Those a bit more south at Eagle Bluffs Conservation Area, Plowboy Bend Conservation Area, or Three Creeks Conservation Area will pick up an additional second. You'll also enjoy 2 minutes and 40 seconds of totality at Columbia Regional Airport and also slightly south of there in Ashland.

Other towns in eastern Missouri at or near the 2 minute and 40 second period of darkness are New Bloomfield, Chamois, St. Clair, Lonedell, De Soto, Olympian Village, and Ste. Genevieve. Even more top the 2½ minute limit: Owensville, Sullivan, Union, Villa Ridge, Richwoods, Hillsboro, Festus, and Herculaneum. Then, the centerline moves through a bit of Illinois before re-entering Missouri.

At 1:19 P.M. CDT, the shadow's midpoint crosses the Mississippi River, which at that location is the state border with Illinois. Once the umbra is fully in the Land of Lincoln, only about a minute passes before totality begins above the point on Earth which will enjoy the longest duration of totality: nearly 2 minutes and 41 seconds in and around Giant City State Park. Murphysboro and Carbondale will be awesome locations, too. So will Chester, Makanda, Pomona, Cobden, Goreville, and Vienna.

The centerline leaves Illinois at its Ohio River border with Kentucky at 1:24:43 P.M. CDT. Totality for Kentucky starts there two minutes earlier and lasts until nearly 1:29 P.M. CDT (depending on your location in the state). The highlight town is Hopkinsville, already billing itself as "the place to be for the eclipse." The celestial circumstances certainly favor it: 2 minutes and 40 seconds of totality with the eclipse's midpoint just to the northwest. My concern is that Hopkinsville, according to the 2010 census, has a population of just 31,577. I hope the city can handle the massive influx of people who surely will venture there for the eclipse. Roads into it are good. Interstate 24 passes just 5 miles south of town, and there's an extension from it that leads right up into Hopkinsville. Also, U.S. Route 68–Bypass is a four-lane, nearly continuous ring around the town.

Other tiny but terrific spots for viewing in Kentucky west of Hopkinsville include Hampton, Burna, Tiline, Fredonia, Eddyville, Kuttawa, Princeton, Dawson Springs, Cadiz, and Cerulean. Favored towns south and east of Hopkinsville include Herndon, Pembroke, Trenton, Elkton, Guthrie, Olmstead, and Adairville. Fort Campbell, near the Tennessee border, enjoys 2½ minutes of totality even though it's well off the centerline, and Franklin sees 2 minutes and 25 seconds of darkness.

After its romp through Kentucky, the centerline crosses the Tennessee border around 1:26 P.M. CDT. I think this state will host the second-highest number of eclipse-watchers, after only Missouri. The main reason: Nashville, which houses 601,000 people according to the 2010 census, with a metro area of nearly 1.6 million souls. Nashville ranks as the largest city fully in the shadow's path. That said, it is not on the centerline. From downtown, viewers will enjoy a second or two under 2 minutes of totality.

Fig. 4.5 Nashville, Tennessee, is the largest city to lie within the region of totality. (Courtesy of Wikimedia Commons)

Die-hard Tennessee shadow-chasers should head northwest to Springfield, north to White House, or northeast to Gallatin where the reward is an extra 41, 42, or 42 seconds of totality, respectively. I hope that by now you realize how important those extra seconds are! Other great destinations include Castalian Springs, Hartsville, Lebanon, Carthage, Granville, Buffalo Valley, Cookeville, Sparta, Crossville, Spring City, Evensville, Ten-Mile, Athens, Sweetwater, Englewood, Madisonville, and Tellico Plains. And a heartfelt "I'm sorry" to Memphis, a city I dearly love. Y'all need to get out of Dodge and make the 200-mile drive to Nashville. Take some blues and bar-b-que with you.

Here's another really important tip: Just past the midpoint of the Volunteer State, the time zone changes to Eastern. Please make a note of this lest your arrival be an hour wide of the mark.

Moving east, the very northeastern tip of North Carolina encounters the centerline from just past 2:35 P.M. EDT until not quite 2:39 P.M. EDT. And although there are only small towns near the centerline, all of these locations will enjoy more than 2½ minutes of totality: Robbinsville, Marble, Andrews, Topton, Hayesville, Franklin, Scaly Mountain, and Highlands.

After North Carolina, the umbral centerline just grazes Georgia. If you need to stay in the Peach State, the best location will be Clayton, which experiences 2 minutes and 37 seconds of totality. Hiawassee and Lakemont rank just behind, with 9 seconds less darkness.

Finally, it's South Carolina's moment in the Sun—I mean shadow. The last of the states the centerline crosses sees totality start at 2:36 P.M. EDT in the northwest part of the state. The city of Greenville lies entirely on the path, though not on the centerline. Residents might think about heading south on Interstate 85 to one of the many small towns nearer the centerline. Doing so will net an extra 20–25 seconds of totality. Some options include Walhalla, Seneca, Clemson, Central, Pendleton, Liberty, Northlake, Centerville, Anderson, Williamston, Belton, and Honea Path. Greenwood and Newberry enjoy 2½ minutes of darkness.

Fig. 4.6 Greenville, South Carolina, enjoys 2 minutes and 10 seconds of totality. (Courtesy of TimothyJ/Wikimedia Commons)

Then it's Columbia's turn. The whole city of 129,000 lies in the shadow's path. So does its metro area, which claims 794,000, according to the 2010 U.S. Census. Columbia is a great location. If you're downtown, you'll revel in 2 minutes and 30 seconds of totality. Those at the city's north end will lose 4 seconds, and people who position themselves at the southern tip will gain 5 seconds. Red Bank, Gaston, and St. Matthews each receive 2 minutes and 36 seconds of darkness. Cross, Moncks Corner, and Huger have 2 seconds less time. If you're in McClellanville, one of the last inhabited spots on land for the eclipse, you'll relish 2 minutes and 32 seconds of totality.

The last land contact of the centerline happens at 2:49:00.5 P.M. EDT at an unmarked location on the waterfront about a mile west of Raccoon Key, the mouth where Raccoon Creek empties into the Atlantic Ocean. Totality here lasts 2 minutes and 35 seconds. After that, the centerline only crosses water. It remains in the Atlantic for another 1 hour 13 minutes and 30 seconds. Then the total solar eclipse of August 21, 2017, passes into the history books.

Fig. 4.7 The center line last touches land south of McClellanville, South Carolina. (Map courtesy of Xavier M. Jubier; data courtesy of Google Imagery/TerraMetrics)

Chapter 5

The Saros Cycle

This section will be pretty complicated, but I included it because some of you may wish to delve more deeply into eclipse relationships. Remember the Saros cycle mentioned previously?

Simply put, the saros cycle is a time period after which nearly identical eclipses repeat. Since the time of the Chaldeans, perhaps as early as the seventh century B.C., astrologers used this period to predict lunar eclipses, though it also applies to solar ones. The Greeks in the second century B.C. also knew of the saros, although it didn't receive that name until 1691, when British astronomer Edmund Halley took the term from a Byzantine lexicon called the *Suda*.

The saros cycle equals 6,585.3211 days. That's how long it takes for four different periods related to the Moon to once again coincide. The first is our satellite's orbital period with respect to the stars, 27.32166 days. Astronomers call this the sidereal period (or sidereal month).

© Springer International Publishing Switzerland 2016
M.E. Bakich, *Your Guide to the 2017 Total Solar Eclipse*, The Patrick Moore
Practical Astronomy Series, DOI 10.1007/978-3-319-27632-8_5

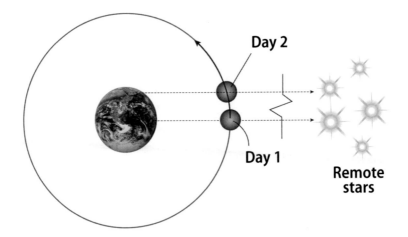

Fig. 5.1 The Moon's sidereal period is the length of time it takes to orbit once with respect to the stars. (Courtesy of Holley Y. Bakich after *Astronomy* magazine: Roen Kelly)

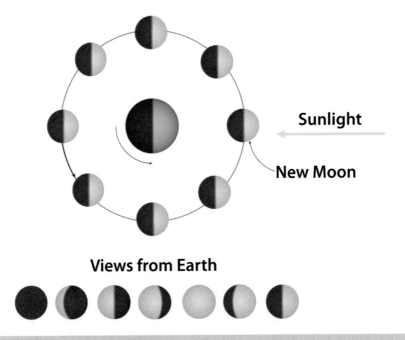

Fig. 5.2 The Moon's synodic period is the length of time it takes to go from one phase to the same phase, as in New Moon to New Moon. (Courtesy of Holley Y. Bakich after *Astronomy* magazine: Roen Kelly)

The second is the synodic period (often called the synodic month or simply the lunar month). This is the length of time, 29.53059 days, it takes for the Moon to go from a particular phase to the next occurrence of that phase. Because we're talking about solar eclipses, we can simplify this to mean that the synodic month is the time between two successive New Moons—the only phase where an eclipse of the Sun can occur.

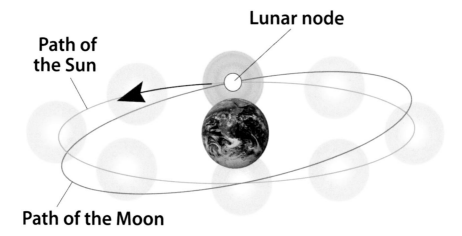

Fig. 5.3 The Moon's draconitic period is the length of time it takes for our satellite to orbit from one node back to the same node. (Courtesy of Holley Y. Bakich after *Astronomy* magazine: Roen Kelly)

We don't experience a solar eclipse every New Moon, however, because our satellite's orbit tilts with respect to Earth's orbit around the Sun. Most of the time the Moon's shadow lies either north or south of Earth, and the three objects do not lie in a straight line.

The plane of the Moon's orbit intersects the ecliptic plane, or the plane of Earth's orbit around the Sun, twice each lunar month at points called nodes. This is the origin of the third of our four periods. A draconitic period (or month) is the time it takes the Moon to go from one node back again to the same node. It equals 27.21222 days.

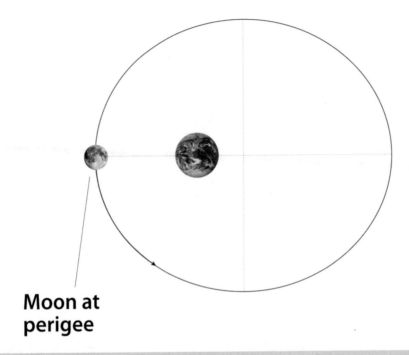

**Moon at
perigee**

Fig. 5.4 The Moon's anomalistic period is the length of time it takes for the Moon to go from one perigee (closest to Earth) to the next. (Courtesy of Holley Y. Bakich after *Astronomy* magazine: Roen Kelly)

The reason two eclipses in different saros cycles have essentially the same durations (including length of totality) is because the Earth–Moon distance is nearly the same for each. If you guessed that this is because of the fourth of the four periods, you're catching on! This one is the anomalistic period (or month), and it equals 27.5545 days. Technically, this is the time between two successive lunar perigees—our satellite's closest approach to Earth.

Now let's put these together to see how they relate. One saros cycle—the next time all four lunar "months" align—equals 241 sidereal periods, which also equals 223 synodic periods, which equals 242 draconic periods, and which also equals 239 anomalistic periods. After one saros cycle, therefore, the positions of the Sun, the Moon, and Earth will be nearly identical. It will be New Moon, our satellite will lie at the same node, and its distance to Earth will be the same.

One more thing will be the same, or nearly so. A saros cycle is only a tad more than 11 days longer than 18 years. In 11 days, Earth travels only 3 percent of its orbit, so its position will be nearly the same. During the same period of time, the Sun's declination (its position with respect to the celestial equator) will change by a maximum of 4° if the date of the eclipse is near an equinox, and hardly at all if the date is near a solstice. What this boils down to is that if you know the date of an eclipse, you can predict that a nearly identical eclipse will occur one saros later with both dates' midday Suns at nearly the same altitude. The second eclipse, however, will occur at a much different place on Earth.

The saros, 6,585.3211 days, is not an integer number. The remainder, 0.3211 day, equals 7 hours 42 minutes and 23 seconds, so each successive eclipse in a saros series happens this much (in many references, you'll see this rounded off to 8 hours) later in the day. For a solar eclipse, therefore, the region of visibility from Earth will shift 115.6° to the west (again, you'll find this rounded to 120°), so you wouldn't be able to observe the two eclipses from the same location, except, perhaps, some of each one's partial phase.

After three saros intervals—54 years and 33 days—the region of visibility has shifted $3 \times 115.6° = 346.8°$, which is less than 14° from a full circle (360°). So the eclipse won't just have the same characteristics as one that occurred 54 years ago, it will occur roughly within an hour of the same time of day. Astronomers call this span a triple saros interval, and you may also see a word derived from the Greek phrase "turn of the wheel": exeligmos (ἐξέλιγμος).

a

b

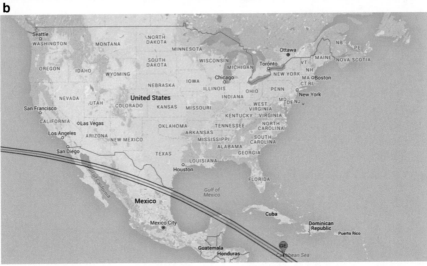

Fig. 5.5 (**a, b**) These two total solar eclipses lie three saros intervals apart. Mid-totality of the August 21, 2017, event occurs at 18h25m29.9s UT. Mid-totality of the September 23, 2071, eclipse occurs just more than an hour earlier at 17h18m09.5s UT. (Map data courtesy of Google/INEGI)

A saros series begins with a group of partial eclipses visible at high geographic latitudes. The first of these partial eclipses occurs when the Sun enters the end of the node. Next, a group of annular or total eclipses is observable from mid-geographic latitudes. Finally, the series ends with another group of partial eclipses, this one near the opposite geographic pole.

Every solar eclipse belongs to such a saros series. Those in a given year, or even a short span of years, belong to different series because they occur much closer than 18 years apart. In 1887, Austrian astronomer Theodor Ritter von Oppolzer published *Canon der Finsternisse* (*Canon of Eclipses*), a compilation of 8,000 solar and 5,200 lunar eclipses that occur between 1207 B.C. and 2161. He found that solar eclipses occur at an average rate of about 238 per century. A quick calculation shows that roughly 42 eclipses occur during a saros period of 18 years. This means that, at any time, approximately 42 different saros series must be active.

It takes between 1,226 and 1,550 years for all the eclipses of a saros series to cross Earth's surface from north to south (or vice versa), meaning that each series contains between 69 and 87 eclipses. Most of them contain 71 or 72 eclipses. Furthermore, between 39 and 59 eclipses in a given series will be central (that is, total, annular, or hybrid). The number you'll encounter most across different series cycles is 43 central eclipses.

The total solar eclipse August 21, 2017, belongs specifically to saros 145. It's the 22nd eclipse in the series, which contains 77. Here's the complete list:

#	Date	Mid eclipse	Type	Location	Totality
1	January 4, 1639	4h56m19s	P	64.6°N 80°E	
2	January 14, 1657	13h08m11s	P	63.7°N 52.7°W	
3	January 25, 1675	21h19m48s	P	62.9°N 175.1°E	
4	February 5, 1693	5h27m09s	P	62.2°N 44.2°E	
5	February 17, 1711	13h30m15s	P	61.6°N 85.4°W	
6	February 27, 1729	21h27m02s	P	61.2°N 146.6°E	
7	March 11, 1747	5h18m08s	P	61°N 20.2°E	
8	March 21, 1765	13h01m45s	P	61°N 104.3°W	
9	April 1, 1783	20h38m39s	P	61°N 132.8°E	
10	April 13, 1801	4h08m06s	P	61.3°N 11.7°E	
11	April 24, 1819	11h31m59s	P	61.7°N 108°W	
12	May 4, 1837	18h48m28s	P	62.3°N 133.9E	
13	May 16, 1855	2h01m12s	P	62.9°N 16.6°E	
14	May 26, 1873	9h08m56s	P	63.7°N 99.6°W	
15	June 6, 1891	16h15m36s	A	74.5°N 163.8°E	0m06s
16	June 17, 1909	23h18m38s	H	82.9°N 123.6°E	0m24s
17	June 29, 1927	6h23m27s	T	78.1°N 73.8°E	0m50s
18	July 9, 1945	13h27m46s	T	70°N 17.2°W	1m15s
19	July 20, 1963	20h36m13s	T	61.7°N 119.6°W	1m40s
20	July 31, 1981	3h46m37s	T	53.3°N 134.1°E	2m02s
21	August 11, 1999	11h04m09s	T	45.1°N 24.3°E	2m23s
22	August 21, 2017	18h26m40s	T	37°N 87.7°W	2m40s
23	September 2, 2035	1h56m46s	T	29.1°N 158°E	2m54s
24	September 12, 2053	9h34m09s	T	21.5°N 41.7°E	3m04s
25	September 23, 2071	17h20m28s	T	14.2°N 76.7°W	3m11s
26	October 4, 2089	1h15m23s	T	7.4°N 162.8°E	3m14s
27	October 16, 2107	9h18m27s	T	1.1°N 40.6°E	3m16s
28	October 26, 2125	17h30m49s	T	4.5°S 83.6°W	3m15s
29	November 7, 2143	1h51m16s	T	9.4°S 150.8°E	3m14s
30	November 17, 2161	10h19m30s	T	13.4°S 23.6°E	3m13s
31	November 28, 2179	18h54m18s	T	16.5°S 104.6°W	3m12s
32	December 9, 2197	3h35m07s	T	18.5°S 126°E	3m13s

#	Date	Mid eclipse	Type	Location	Totality
33	December 21, 2215	12h20m08s	T	19.5°S 4.1°W	3m14s
34	December 31, 2233	21h07m37s	T	19.5°S 134.7°W	3m18s
35	January 12, 2252	5h57m05s	T	18.5°S 94°E	3m23s
36	January 22, 2270	14h46m29s	T	16.7°S 37.3°W	3m29s
37	February 2, 2288	23h33m47s	T	14.2°S 168.4°W	3m39s
38	February 14, 2306	8h17m49s	T	11.3°S 61°E	3m49s
39	February 25, 2324	16h57m32s	T	8.1°S 68.6°W	4m02s
40	March 8, 2342	1h32m14s	T	4.9°S 162.9E	4m16s
41	March 18, 2360	9h59m22s	T	1.8°S 36.4°E	4m33s
42	March 29, 2378	18h20m23s	T	1.1°N 88.6°W	4m51s
43	April 9, 2396	2h33m17s	T	3.4°N 148.7°E	5m12s
44	April 20, 2414	10h39m39s	T	5°N 27.7°E	5m33s
45	April 30, 2432	18h37m31s	T	5.8°N 91°W	5m56s
46	May 12, 2450	2h29m44s	T	5.6°N 151.7°E	6m19s
47	May 22, 2468	10h15m11s	T	4.2°N 36°E	6m41s
48	June 2, 2486	17h55m28s	T	1.8°N 78.7°W	6m59s
49	June 14, 2504	1h31m03s	T	1.9°S 167.3°E	7m10s
50	June 25, 2522	9h03m45s	T	6.6°S 53.5°E	7m12s
51	July 5, 2540	16h34m26s	T	12.4°S 60.6°W	7m04s
52	July 17, 2558	0h03m14s	T	19.2°S 175°W	6m43s
53	July 27, 2576	7h32m31s	T	26.9°S 69.5°E	6m12s
54	August 7, 2594	15h02m42s	T	35.6°S 47.4°W	5m32s
55	August 18, 2612	22h35m27s	T	45.2°S 166.8°W	4m45s
56	August 30, 2630	6h10m52s	T	56.1°S 68.9E	3m53s
57	September 9, 2648	13h51m23s	T	70.1°S 79.1°W	2m48s
58	September 20, 2666	21h37m07s	P	72.2°S 136°E	
59	October 1, 2684	5h28m03s	P	72.1°S 4.5°E	
60	October 13, 2702	13h25m52s	P	71.8°S 128.5°W	
61	October 23, 2720	21h30m13s	P	71.2°S 97.2°E	
62	November 4, 2738	5h42m27s	P	70.5°S 38.5°W	
63	November 14, 2756	13h59m51s	P	69.6°S 174.9°W	
64	November 25, 2774	22h25m15s	P	68.6°S 47.4°E	
65	December 6, 2792	6h55m37s	P	67.6°S 91°W	
66	December 17, 2810	15h31m45s	P	66.5°S 129.7°E	
67	December 28, 2828	0h10m20s	P	65.4°S 9.7°W	
68	January 8, 2847	8h52m24s	P	64.5°S 149.6°W	
69	January 18, 2865	17h34m21s	P	63.6°S 70.9E	
70	January 30, 2883	2h15m42s	P	62.8°S 68.1°W	
71	February 10, 2901	10h54m17s	P	62.1°S 153.8°E	
72	February 21, 2919	19h29m46s	P	61.7°S 16.6°E	
73	March 4, 2937	3h58m34s	P	61.3°S 118.8°W	
74	March 15, 2955	12h21m15s	P	61.2°S 107.5°E	
75	March 25, 2973	20h35m44s	P	61.2°S 24.2°W	
76	April 6, 2991	4h43m03s	P	61.4°S 154.2°W	
77	April 17, 3009	12h39m22s	P	61.7°S 78.6°E	

Key: A=Annular; H=Hybrid; P=Partial; T=Total; Times are given in Universal Time; Durations of totality are given for the maximum along the shadow path.

Chapter 6

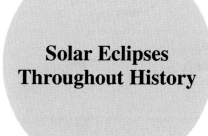

Solar Eclipses Throughout History

Eclipses have occurred throughout human history, creating awe and fear whenever people observed them. They still create awe, if not fear. It's fun to read about past eclipses, so I've collected accounts of some recorded by a number of chroniclers who were not astronomers, although some had interest in what was happening in the sky. Today, "historical" eclipses refer to those before the invention of the telescope in 1608. Every eclipse after that date, then, must be a "modern" one, at least in the astronomical sense.

Because this chapter deals with past eclipses, I chose to illustrate it with drawings and photographs from my collection of 19th-century books. Unfortunately, photography isn't all that old. The first eclipse photograph dates from 1851, so no earlier exact representation of these events exists. So, how accurate are the sketches? Actually, better than we might think.

Astronomers—especially those who lived in the pre-photography "visual" days—carefully trained their eyes to pick out minute detail. Those that attempted to record the Sun's corona on paper generally had steady hands. The ultimate judges, of course, were contemporary astronomers who had witnessed the same events. Most of the sketches that made it into print, therefore, had been critiqued beforehand. For that reason, we can be fairly certain that those we are seeing convey the Sun's appearance well.

© Springer International Publishing Switzerland 2016
M.E. Bakich, *Your Guide to the 2017 Total Solar Eclipse*, The Patrick Moore
Practical Astronomy Series, DOI 10.1007/978-3-319-27632-8_6

Ancient Times

Let's start our recounting with the earliest recorded eclipse of the Sun, one mentioned in the ancient Chinese tome, *Shou-Ching* (or, *Chou-King*). The text relating to the eclipse translates as, "On the first day of the last month of autumn, the Sun and Moon did not meet harmoniously in Fang." For quite some time after much study, astronomers and historians concluded the eclipse described occurred either in the year 2136 B.C. or 2128 B.C.

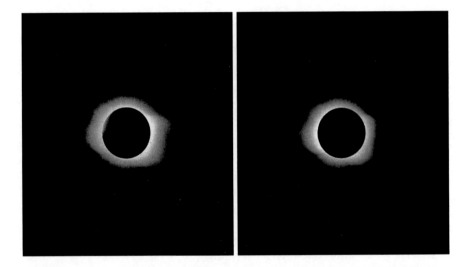

Fig. 6.1 (a, b) The 0.5-second exposure on the left was taken immediately after the beginning of totality during the January 1, 1889, total solar eclipse. The 3-second exposure on the right was taken about 45 seconds later. (Courtesy of M. E. Bakich library)

Helping astronomers narrow the field is the fact that "Fang" refers to a particular part of the sky. It includes the area bounded by the stars Beta (β), Delta (δ), Pi (π), and Rho (ρ) in the constellation Scorpius, a few faint stars in Libra and Ophiuchus to the north, and several others in Lupus to the south.

This eclipse happened during the reign of Chung-K'ang, the fourth Emperor of the Hsia Dynasty. The event is interesting not only for its antiquity, but also for the judgment against two royal astrologers (the astronomers of their day), Hsi (or Hi) and Ho, of the period. They were taken by surprise when it occurred and did not perform the customary rites, which included the shooting of arrows and the beating of drums, gongs, and anything else that could make noise with the purpose of delivering the Sun from the monster threatening to devour it.

According to the story, Hi and Ho were drunk and incapable of performing their duties. Turmoil ensued when the royal court determined the gods were angry. To appease the gods, and to punish the astrologers for their misconduct, Hi and Ho were put to death. This punishment may have been a bit excessive. Why? Because the eclipse in question wasn't even a total one, only a partial.

An anonymous verse appeared thousands of years after the incident:

Here lie the bodies of Ho and Hi,
Whose fate, though sad, was risible.
Being hanged because they could not spy
Th'eclipse, which was invisible.

We now know that the eclipse in question occurred October 22, 2137 b.c. The problem in setting the date originally arose from the fact that there was an uncertainty of 108 years in the date when Emperor Chung-K'ang ascended the throne. Within that span, there were 14 possible years when an eclipse of the Sun in Fang could have occurred. Researchers further reduced that number by determining which eclipse could have been seen from the Emperor's capital.

The Chinese also are responsible for the second-earliest eclipse of the Sun of which we have a record. Confucius (K'ung Fu-tzu) relates that one took place during the reign of the Emperor Yew-Wang. He reigned between 781 b.c. and 771 b.c. The most likely fit during this time is the eclipse of June 4, 780 b.c.

Fig. 6.2 The sketch at the top is from the July 29, 1878, total solar eclipse. The one below it was made during the total solar eclipse of December 12, 1871. (Courtesy of M. E. Bakich library)

Confucius also wrote five major Chinese historical works, one of which, the *Chūnqiū (The Spring and Autumn Annals)*, gives an account of 36 solar eclipses observed in China, the first of which was on February 22, 720 B.C., and the last on July 22, 495 B.C. John Williams, assistant secretary of the Royal Astronomical Society in London, presented a paper to that August body about this work in 1863. The accounts he drew from it were little more than brief mentions of the events. Here's one example: "In the 58th year of the 32nd cycle in the 51st year of the Emperor King-Wang, of the Chow Dynasty, the 3rd year of Yin-Kung, Prince of Loo, in the spring, the second moon, on the day called Kea-Tsze, there was an eclipse of the Sun." This 58th year of the 32nd cycle works out to 720 B.C. in the western way of reckoning dates from that time.

One other early Chinese work has provided confirmation of solar eclipses that were observed from that empire. The *Tsze che tung kēen kang muh (Abridgement of the Mirror of History)*, first published in 1473, is a work of more than 50 volumes containing a summary of Chinese history from the earliest times to the end of the Yuen Dynasty in 1368. It includes brief accounts of solar eclipses, bright meteors and meteor showers, and visible comets, but, strangely enough, not a single mention of a lunar eclipse.

Of course, eclipses were observed worldwide at and before the dates noted. It's just that the ancient Chinese tended to chronicle events in the sky more thoroughly.

In classical Greece, historians began to record eclipses that occurred starting in the sixth century B.C. Plutarch (A.D. 46–120), in his *Life of Romulus*, refers to a remarkable incident connected to the legendary figure's death: "The air on that occasion was suddenly convulsed and altered in a wonderful manner, for the light of the Sun failed, and they were involved in an astonishing darkness, attended on every side with dreadful thunderings and tempestuous winds." An annular eclipse did occur near the supposed date of Romulus' death in 716 B.C. On December 10 of that year, the path of annularity swept through northern Africa. Greece would have experienced one-third of the solar disc covered by the Moon.

Apparently the Greek lyric poet Archilochus (c. 680–c. 645 B.C.) experienced a solar eclipse during his lifetime. Only fragments of his writings remain, but one relates, "Zeus the father of the Olympic Gods turned midday into night hiding the light of the dazzling sun; an overwhelming dread fell upon men." It's probable that he refers to the total eclipse April 6, 648 B.C. In southern Greece, the populace would have experienced nearly 5 minutes of totality with the peak occurring around 10 A.M.

Fig. 6.3 This photograph shows the viewing site set up in Mina Bronces, Chile, by astronomers from Lick Observatory, for the April 16, 1893, total solar eclipse. (Courtesy of M. E. Bakich library)

Some six decades later, a total solar eclipse that occurred May 28, 585 B.C., might have proved instrumental in ending a war between the Lydians and the Medes. And just as interesting, some accounts state that the antiquarian scientist Thales of Miletus (c. 624–c. 546 B.C.) predicted this eclipse.

Herodotus (c. 484–c. 425 B.C.) in Book I of his *Histories* describes a war between the two cultures and gives the circumstances which led to it ending: "As the balance had not inclined in favor of either nation, another engagement took place in the sixth year of the war, in the course of which, just as the battle was growing warm, day was suddenly turned into night. This event had been foretold to the Ionians by Thales of Miletus, who predicted it for the very year in which it actually took place. When the Lydians and Medes observed the change they ceased fighting, and were alike anxious to conclude peace." To cement the bond, a twofold marriage took place. Herodotus adds, "For without some strong bond, there is little security to be found in men's covenants."

The exact date of this eclipse was the subject of scholarly debate for some years. In the end, no less a figure than George Biddell Airy, England's seventh Astronomer Royal, settled the date after exhaustive research. To criticism that Thales could not have predicted the event, Airy wrote, "I think it not at all improbable that the eclipse was so predicted, and there is one easy way, and only one of predicting it—namely, by the *Saros*, or period of 18 years, 10 days, 8 hours nearly. By use of this period an evening eclipse may be predicted from a morning eclipse, but a morning eclipse can rarely be predicted from an evening eclipse (as the interval of eight hours after an evening eclipse will generally throw the eclipse at the end of the *Saros* into the hours of night). The evening eclipse, therefore, of B.C. 585, May 28, which I adopt as being most certainly the eclipse of Thales, might be predicted from the morning eclipse of B.C. 603, May 17 ... No other of the eclipses discussed by Baily and Oltmanns present the same facility for prediction."

Almost exactly one century later, a solar eclipse affected the decision of another leader. After the battle of Thermopylae (made famous by the movie *300*), the Peloponnesian Greeks commenced to fortify the isthmus of Corinth with the view of defending it with their small army against the invading Persian host led by Xerxes. The Spartan troops were under the command of Cleombrotus, the brother of Leonidas, the hero of Thermopylae. Cleombrotus had been consulting the oracle at Sparta, and the historian Herodotus writes that, "while he was offering sacrifice to know if he should march out against the Persian, the Sun was suddenly darkened in mid-sky."

Fig. 6.4 Alfred Brothers of Manchester, England, created this sketch from a photograph he took of the corona during the December 22, 1870, total solar eclipse at Syracuse, Sicily. (Courtesy of M. E. Bakich library)

This event so frightened Cleombrotus that he returned home with his forces. Because Herodotus writing isn't exhaustive about this point, scholars don't know if Cleombrotus returned to Sparta in the autumn of the year of the battle of Salamis or in the spring of the following year, when the battle of Plataea was fought. The best guess is the former because it coincides with the solar eclipse of October 2, 480 B.C. The Sun, roughly 60 percent obscured, would have been high in the heavens near midday.

Another military commander dealt with a terrifying eclipse in a totally different way. The historian Thucydides (c. 460–c. 400 B.C.) writes that during the first year of the Peloponnesian War, "in the same summer, at the beginning of a new lunar month (at which time alone the phenomenon seems possible) the Sun was eclipsed after mid-day, and became full again after it had assumed a crescent form and after some of the stars had shone out."

Astronomers recognize August 3, 431 B.C. as the date of this eclipse. From Greece, the Moon covered more than 80 percent of the Sun. Venus was 20° and Jupiter 43° from the action, so those were probably the "stars" seen.

This eclipse nearly stopped the Athenian attack on the Lacedaemonians. The Athenian sailors were terrified, but their commander, Pericles, had an idea. According to Plutarch, writing in his *Life of Pericles*, "The whole fleet was in readiness, and Pericles on board his own galley, when there happened an eclipse of the Sun. The sudden darkness was looked upon as an unfavorable omen, and threw the sailors into the greatest consternation. Pericles observing that the pilot was much astonished and perplexed, took his cloak, and having covered his [the pilot's] eyes with it, asked him if he found anything terrible in that, or considered it as a bad presage? Upon his answering in the negative, he said, 'Where is the difference, then between this and the other, except that something bigger than my cloak causes the eclipse?'"

The ancient eclipses I just described, although interesting in their own right, were of value to early (especially 19th-century) astronomers in helping to establish exactly how the Moon moved in its orbit. But even a century ago, some people warned against relying too much on the supposed accuracy of those accounts.

For example, in 1875, Canadian-American astronomer Simon Newcomb (1835–1909) wrote, "The first difficulty is to be reasonably sure that a total eclipse was really the phenomenon observed. Many of the statements supposed to refer to total eclipses are so vague that they may be referred to other less rare phenomena. It must never be forgotten that we are dealing with an age when accurate observations and descriptions of natural phenomena were unknown, and when mankind was subject to be imposed upon by imaginary wonders and prodigies. The circumstance, which we should regard as most unequivocally marking a total eclipse, is the visibility of stars during the darkness. But even this can scarcely be regarded as conclusive, because Venus may be seen when there is no eclipse, and may be quite conspicuous in an annular or a considerable partial eclipse. The exaggeration of a single object into a plural is in general very easy. Another difficulty is to be sure of the locality where the eclipse was total. It is commonly assumed that the description necessarily refers to something seen where the writer flourished, or where he locates his story. It seems to me that this cannot be safely done unless the statement is made in connection with some battle or military movement, in which case we may presume the phenomena to have been seen by the army."

The Christian Era—Millennium #1

On November 24, 29 A.D., a solar eclipse sometimes called the "eclipse of Phlegon" (second century A.D.) occurred. It got this name because the historian Origen (182–254 A.D.) recorded Phlegon's account of it. In Book II, chapter 23 of Phlegon's *Contra Celsum* (*Against Celsus*), he writes of an eclipse accompanied by earthquakes during the reign of Tiberius: that there was "the greatest eclipse of the sun" and also that "it became night in the sixth hour of the day [i.e., noon] so that stars even appeared in the heavens. There was a great earthquake in Bithynia, and many things were overturned in Nicaea."

Fig. 6.5 This spectacular sketch represents the view of the Sun's corona through a 4-inch telescope during the August 7, 1869, total solar eclipse. (Courtesy of M. E. Bakich library)

This was the only solar eclipse visible from Jerusalem during the years usually given as those of Christ's public ministry. Many writers since have associated this eclipse with the darkness described in the Bible that occurred at the Crucifixion. Unfortunately, the obscuration of the Sun by the Moon during this eclipse at Jerusalem was only 80 percent—far from total. Also, the generally recorded date of the Crucifixion, April 3, 33 A.D., coincided with a Full Moon. Interestingly, that Full Moon was eclipsed by Earth's shadow, but the darker umbra passed from the Moon's face before it rose at Jerusalem.

The next eclipse of note makes my "Top 10 Eclipses of All Time." (And just so you know, the first eclipse on this list was the one "missed" by Hsi and Ho.) It occurred March 20, 71, and what makes it famous is that the Greek philosopher Plutarch became the first to describe the corona, the thin outer atmosphere of the Sun, in his book *De Facie in Orbe Lunae* (On the face which appears in the Moon). Plutarch, through a character in the book named Lucien, says, "Even if the moon, however, does sometimes cover the sun entirely, the eclipse does not have duration or extension; but a kind of light is visible about the rim which keeps the shadow from being profound and absolute."

For the next thousand years, the accuracy of surviving eclipse reports actually declined. Compared to Greek skywatchers, the ecclesiastical historians and monks who jotted down meager descriptions were poor chroniclers. Indeed, the term "Dark Ages," as applied to this time (and specifically by many historians to the period from the fifth to the tenth century) certainly has no relation to solar eclipses. The only thing seemingly "dark" about them was the lack of details communicated.

Still, in the first millennium A.D., writers recorded noted total solar eclipses in various publications. The Roman Julius Capitolinus, along with five other authors, wrote *Historia Augusta* (*The Augustan History*), which was a collection of biographies. He described the total solar eclipse of April 12, 237, as being so great "that people thought it was night, and nothing could be done without lights." Totality occurred just to the north of present-day Italy, and lasted 3 minutes and 8 seconds at the current location of Zurich (which the Romans founded about 2,000 years ago, calling it Turicum).

Another of my picks for Top 10 eclipses goes to the total solar eclipse of July 17, 334. The Latin writer and astrologer Julius Firmicus Maternus gave the earliest description of a prominence during this event, which he observed from Sicily. Interestingly, this event was an annular eclipse and not a total one.

The total solar eclipse of July 19, 418, is a notable one in the history of such events. The path of totality crossed Spain, Italy (the center line passed just north of Rome where totality would have lasted some 3½ minutes), Turkey, Iran, and India. But it's not the path or the length of this eclipse that sets it apart. Rather, it's what was seen with it.

I also consider this one of the Top 10 eclipses because of the following discovery. Turkish-born Church historian Philostorgius (368–439) described the appearance of a comet during totality of this eclipse. In Book XII of the *Epitome*

Historiae Ecclesiasticae, he wrote, "When Theodosius had reached adolescence, on the nineteenth of July at about the eighth hour, the sun was so completely eclipsed that stars appeared. And such a drought followed this event that there was everywhere an unusually high number of deaths of human beings and animals.

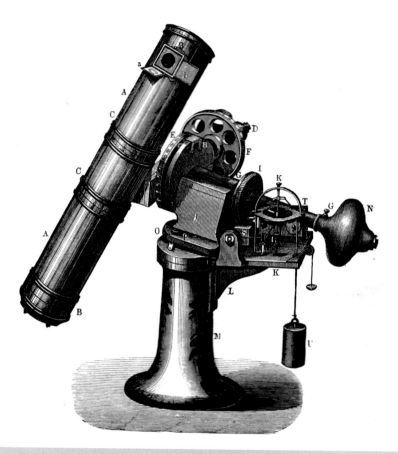

Fig. 6.6 This photographic telescope, made by Browning for the Indian government to take pictures of the eclipse August 18, 1868, has the three components deemed necessary by 19th-century astronomers: an astronomical telescope, a driving clock, and a photographic apparatus. (Courtesy of M. E. Bakich library)

"There appeared in the sky with the sun while in eclipse a cone-shaped light, which some out of ignorance called a comet. But it showed none of the features of a comet. For the light did not form a tail, nor was it at all like a star; rather, it resembled a great lamp-flame appearing on its own, with no star under it to form a wick for it. Its movement was also different. It began where the sun rises at the equinox, from there passed over the last star in the Bear's tail, and went on slowly westward. But when it had traversed the sky, it disappeared, having taken more than four months to make its journey."

Other European texts recorded a comet appearing in the eastern sky after the eclipse. A Byzantine account states it was visible for 7 months.

The Chinese recorded a comet in the Big Dipper in June of 418 and wrote about it again when it appeared in Leo and Virgo in September. At that point, it displayed a tail more than 100° long.

Although Philostorgius discounted this object as being a comet, it certainly was one, so historians have credited him with the first description of a comet seen during totality.

The first description of an English record of an eclipse observed in England didn't appear until well into the sixth century. American author Mabel Loomis Todd, writing in *Total Eclipses of the Sun*, explains the deficiency thusly: "The accounts, however, are greatly confused and uncertain, as would perhaps be natural fully 60 years before the advent of St. Augustine, and when Britain was helplessly harassed with its continual struggle in the fierce hands of West Saxons and East Saxons, of Picts and conquering Angles. Men have little time to record celestial happenings clearly, much less to indulge in scientific comment and theorizing upon natural phenomena, when the history of a nation sways to and fro with the tide of battle, and what is gained today may be fatally lost tomorrow. And so there is little said about this eclipse, and that little is more vague and uncertain even than the monotonous plaints of Gildas—the one writer whom Britain has left us, in his meagre accounts of the conquest of Kent, and the forsaken walls and violated shrines of this early epoch."

The description, sparse as it is, comes from volume II of the *Anglo-Saxon Chronicle*, a collection of annals in Old English begun around 890 and updated through the middle of the 12th century: "In this year the Sun was eclipsed 14 days before the Calends of March from early morning till 9 A.M." Although this eclipse was total, the Moon's umbra passing through northern Africa, Turkey, and Russia, at London, Luna would have covered only 60 percent of the Sun's disk.

Millennium #2

One of the most celebrated total solar eclipses of medieval times occurred August 2, 1133, visible in Scotland as one with a totality nearly 4½ minutes in length. This eclipse demonstrates how people can wrongly associate such a celestial spectacle with an earthly event.

Fig. 6.7 Captain G. L. Tupman of the U.S. Naval Observatory made this sketch of the December 22, 1870, eclipse at Syracuse, Sicily, while observing through a 1.2-inch refractor. (Courtesy of M. E. Bakich library)

The noted 12th-century historian William of Malmesbury correctly related this eclipse to the departure of Henry I the last time he went to Normandy: "The elements manifested their sorrow at this great man's last departure from England. For the Sun on that day at the 6th hour shrouded his glorious face, as the poets say, in hideous darkness agitating the hearts of men by an eclipse; and on the 6th day of the week early in the morning there was so great an earthquake that the ground appeared absolutely to sink down; an horrid noise being first heard beneath the surface."

The *Anglo-Saxon Chronicle* also refers to this eclipse, although that source wrongly gives the year as 1135: "In this year King Henry went over sea at Lammas, and the second day as he lay and slept on the ship the day darkened over all lands; and the Sun became as it were a three-night-old Moon, and the stars about it at mid-day. Men were greatly wonder-stricken and affrighted, and said that a great thing should come hereafter. So it did, for the same year the king died on the following day after St. Andrew's Mass-day, Dec. 2, in Normandy." The king did die in 1135, but the eclipse occurred 2 years earlier.

When she wrote about this discrepancy, Mabel Loomis Todd stated, "So Henry must have died in 1133, which he *did not*; or else there must have been an eclipse in 1135, which there *was not*. But this is not the only labyrinth into which chronology and old eclipses, imagination, and computation, lead the unwary searcher."

One of the most famous eclipses in the Middle Ages occurred June 17, 1433. Inhabitants of Scotland remembered this event as the "Black Hour." Indeed, it must have been quite a sight. Totality above that country lasted between 4 minutes and 22 seconds and 4 minutes and 26 seconds. Two later Scottish total solar eclipses also received nicknames: "Black Saturday" refers to the eclipse March 7, 1598, whose duration of totality lasted for 1 minute at Edinburgh. "Mirk Monday" is the name given to the eclipse that occurred April 8, 1652. The center line of this eclipse passed through Edinburgh, where residents experienced 2 minutes and 46 seconds of totality.

In 1544, a hybrid solar eclipse occurred with a totality that lasted a scant 16.3 seconds. The chief interest in this event was that it was one of the first eclipses observed by a professed (if not professional) astronomer. The Dutch mathematician Gemma Frisius saw it at Louvain (Leuven), Belgium, where the Moon obscured 95 percent of the Sun's disk. Afterward, Gemma Frisius published probably the first illustration of a camera obscura. It shows an image of a solar eclipse projected onto the wall of a room.

Years later, German astronomer Johannes Kepler, who wasn't even born until 1571, said that the day became dark like the twilight of evening and that the birds, which from the break of day had been singing, became mute. Kepler, by the way, first used the term "camera obscura" (from the Latin *camera*, [vaulted] chamber or room and *obscura*, darkened).

This famous astronomer also used a camera obscura to observe the total solar eclipse July 10, 1600, although in his location this event would have been a partial eclipse with a maximum coverage of 49 percent of the Sun's disk. He made this observation in the marketplace in Graz, Austria.

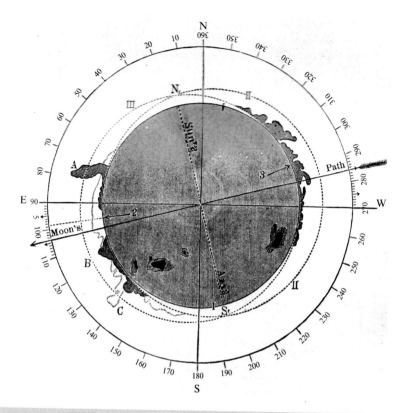

Fig. 6.8 Indian born Scottish astronomer James Francis Tennant collected many photographs of the August 18, 1868, total solar eclipse and combined them into this drawing. (Courtesy of M. E. Bakich library)

Sometimes correct observations during a total solar eclipse can lead to totally wrong conclusions. Consider the eclipse May 12, 1706, visible as a partial eclipse in England but total in Switzerland. A certain Captain Henry Stannyan, observing from Berne, sent a description to John Flamsteed, the first Astronomer Royal of England: "That the Sun was totally darkened there for four and a half minutes of time; that a fixed star and a planet appeared very bright; and that his getting out of his eclipse was preceded by a blood-red streak of light from its left limb, which continued not longer than six or seven seconds of time; then part of the Sun's disk appeared all of a sudden as bright as Venus was ever seen in the night; nay, brighter; and in that very instant gave a light and shadow to things as strong as the Moon uses to do."

When Flamsteed remarked about Stannyan's comments to the Royal Society, he said, "The Captain is the first man I ever heard of that took notice of a red streak preceding the emersion of the Sun's body from a total eclipse, and I take notice of it to you, because it infers that the Moon has an atmosphere; and its short continuance, if only six or seven seconds' of time, tells us that its height was not more than five or six hundredths part of her diameter." Ouch! Flamsteed was a great astronomer, but he seriously erred in this case.

Now for another of my Top 10 eclipses. This one occurred May 22, 1724. Although Plutarch had described the corona some 17 centuries earlier, it was the Spanish astronomer José Joaquín de Ferrer who gave the phenomenon its name after viewing this total solar eclipse from Kinderhook, New York. *Corona* is Latin for "crown." He also correctly surmised that the corona was part of the Sun rather than the Moon because of its size.

The next two total solar eclipses of note were unusual for the 18th century because both were observed at sea, and neither was an eclipse expedition, per se. On February 9, 1766, officers aboard the French man-of-war *Comte d'Artois* observed a totality lasting 53 seconds while they were cruising the Indian Ocean. They reported seeing a luminous ring around the Moon that had four remarkable expansions at a [n angular] distance of 90° from each other.

Then, 12 years later, on June 24, 1778, Spanish Admiral Don Antonio Ulloa was passing from the Azores to Cape St. Vincent when he observed a totality lasting 4 minutes. The luminous corona appeared beautiful, and observers saw coronal streamers reaching a distance that equaled the Moon's diameter. One interesting note states that before totality became conspicuous, first- and second-magnitude stars were visible, but at mid-eclipse only first-magnitude stars could be seen. Ulloa wrote, "The darkness was such that persons who were asleep and happened to wake, thought that they had slept the whole evening and only waked when the night was pretty far advanced. The fowls, birds, and other animals on board took their usual positions for sleeping, as if it had been night."

Just one more minor historical note related to this event. This was the first total solar eclipse observed in the U.S. after it became a country. Inhabitants in four of the original 13 "colonies" (Virginia, North Carolina, South Carolina, and Georgia) experienced totalities between 4 minutes and 51 seconds and 5 minutes and 21 seconds.

The 19th Century

Eclipse observations prior to 1801 were, except in rare cases, made by essentially unskilled persons with no clear ideas as to what they should look for or what they might expect to see. Things improved a bit during the 18th century with astronomers like Edmund Halley and Sir William Herschel, but it wasn't until the 19th century — and if I'm honest, the middle of that hundred-year span — that researchers began to observe eclipses under circumstances that would let them extract serious science.

Fig. 6.9 These sketches from British astronomer Charles Piazzi Smyth show the darkness during the total eclipse July 28, 1851. (Courtesy of M. E. Bakich library)

The total solar eclipse of June 16, 1806, was visible throughout a huge swath of the U.S. The Moon's shadow entered the country in the Desert Southwest and passed slightly north of Boston before heading out into the Atlantic Ocean. Observers on the center line from positions north of Wichita, Kansas, through the Bay State experienced more than 4 minutes of totality. This event had a profound effect on many American astronomers including William Cranch Bond, who became the first director of Harvard College Observatory.

One of the eclipses I rate as in the Top 10 ever occurred May 15, 1836, and, believe it or not, it was an annular eclipse. In northern England and southern Scotland, observers saw some 4½ minutes of annularity. English astronomer

Francis Baily, viewing from Jedburgh in Roxburghshire, provided the first explanation of why we see beads of sunlight just before and just after totality during solar eclipses (and also, obviously, during some annular eclipses). He explained that it is because we are viewing the Moon's irregular surface, especially hills and valleys, which alternately block or allow sunlight to peek through to our eyes. To commemorate this discovery, the phenomenon is now known as Baily's beads.

After this eclipse, French astronomer François Arago wrote in the 1846 edition of *L'Annuaire* a wonderful account of the effect it had on the local populace. As we look forward to the August 21, 2017 eclipse, those of us who will be in large groups should take note. It's quite likely that the response of the crowd we're with will be the same as the one Arago viewed the eclipse with in 1836: "The hour of the commencement of the eclipse drew nigh. More than twenty thousand persons, with smoked glasses in their hands, were examining the radiant globe projected upon an azure sky. Although armed with our powerful telescopes, we had hardly begun to discern the small notch on the western limb of the Sun, when an immense exclamation, formed by the blending together of twenty thousand different voices, announced to us that we had anticipated by only a few seconds the observation made with the unaided eye by twenty thousand astronomers equipped for the occasion, whose first essay this was. A lively curiosity, a spirit of emulation, the desire of not being outdone, had the privilege of giving to the natural vision an unusual power of penetration. During the interval that elapsed between this moment and the almost total disappearance of the Sun we remarked nothing worthy of relation in the countenances of so many spectators. But when the Sun, reduced to a very narrow filament, began to throw upon the horizon only a very feeble light, a sort of uneasiness seized upon all; every person felt a desire to communicate his impressions to those around him. Hence arose a deep murmur, resembling that sent forth by the distant ocean after a tempest. The hum of voices increased in intensity as the solar crescent grew more slender; at length the crescent disappeared, darkness suddenly succeeded light, and an absolute silence marked this phase of the eclipse with as great precision as did the pendulum of our astronomical clock. The phenomenon in its magnificence had triumphed over the petulance of youth, over the levity which certain persons assume as a sign of superiority, over the noisy indifference of which soldiers usually make profession. A profound stillness also reigned in the air; the birds had ceased to sing. After an interval of solemn expectation, which lasted about two minutes, transports of joy, shouts of enthusiastic applause, saluted with the same accord, the same spontaneous feeling, the first reappearance of the rays of the Sun. To a condition of melancholy produced by sentiments of an indefinable nature there succeeded a lively and intelligible feeling of satisfaction which no one sought to escape from or moderate the impulses of. To the majority of the public the phenomenon had arrived at its term. The other phases of the eclipse had few attentive spectators beyond the persons devoted especially to astronomical pursuits."

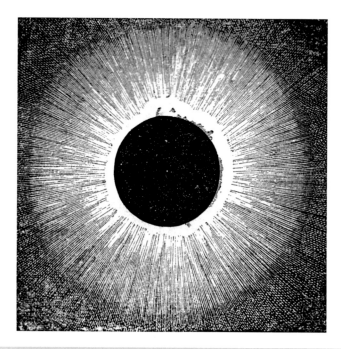

Fig. 6.10 This detailed sketch by English astronomer William Rutter Dawes reveals prominences seen during the total solar eclipse July 28, 1851. (Courtesy of M. E. Bakich library)

English astronomy popularizer George F. Chambers wrote that the total eclipse of July 28, 1851, was the first that was the subject of an "eclipse expedition." Totality occurred in Norway and Sweden (as well as other locations), and great numbers of astronomers from all parts of Europe flocked to those two countries. Among the noteworthy scientists from England were Astronomy Royal George Biddell Airy, John Russell Hind, and William Lassell.

This event makes my Top 10 list because a Russian photographer named Berkowski (nobody knows his first name) took the first successful image of totality during this eclipse. He mounted a 2.4-inch refractor atop one of the instruments at the Royal Observatory in Königsberg, Prussia (now Kalingrad, Russia), and exposed a daguerreotype plate for 84 seconds.

Apparently, Top 10 entries group together. Another one on my list occurred August 18, 1868. During totality, French astronomer Pierre Jules César Janssen and British astronomer J. Norman Lockyer independently discover an unknown line in the Sun's spectrum. Lockyer proposed that the new line was due to the presence of a hitherto undiscovered element, which he named "helium" because in mythology Helios was the name of the Greek Sun god. This element—the second most abundant in the cosmos—would not be found on Earth for another 27 years!

Fig. 6.11 Astronomers from the U.S. Naval Observatory set up this observatory in Des Moines, Iowa, for the August 7, 1869 eclipse. (Courtesy of M. E. Bakich library)

An eclipse that proved to astronomers the value of high-altitude observing locations occurred July 29, 1878. This one crossed the U.S. from the northwest to the southeast, but many astronomers set up observation posts at elevated locations in the Rocky Mountains.

Atop the 14,115-foot-high (4,302 meters) summit of Pike's Peak, a group of skilled observers led by American astronomer Samuel Pierpont Langley made some incredible sightings. Comparing their views of coronal streamers to a party of astronomers led by English astronomer Simon Newcomb at an elevation of 8,000 feet (2,440 meters), showed a marked improvement caused by the clarity of the air at the higher location.

And here's a follow-up note that I can hardly believe. Astronomers in Langley's group claim that the corona remained visible for fully 4 minutes after totality!

The next eclipse of note occurred May 17, 1882. And the noteworthy thing about it was that after English astronomer Arthur Schuster had developed the photographic plates he took during totality, he discovered a comet within the corona.

Astronomers made many preparations to observe the total solar eclipse of August 19, 1887, which crossed a vast stretch of Russia as well as parts of China and Japan. Unfortunately, clouds prevailed in most locations. One observer who had a clear view was the Russian chemist Dmitri Mendeleev, who rose to an altitude of some two miles in a hot-air balloon. His reward was a spectacular view of the solar corona, but because of miscommunication, the assistant that should have accompanied him remained on Earth. So, much of the time that Mendeleev would have spent in scientific study of the eclipse was taken up because he had to manage the balloon alone.

There is one eclipse, however, that tops any of the ones I have presented in this chapter. It occurred May 29, 1919, crossed the width of both South American and central Africa, and had one of the longest durations of totality in the 20th century, 6 minutes and 51 seconds. But these facts are not what make this eclipse famous.

In 1916, German-born physicist Albert Einstein published the general theory of relativity. In it, Einstein said space could be curved by of the influence of the gravity of any body with mass. To prove this, English astrophysicist Arthur Eddington proposed a test during which observers would photograph the Sun during a solar eclipse's totality to see if the Sun's gravity made any stars visible that would not normally be seen.

Indeed, during the 1919 eclipse, Eddington led an expedition to the island of Príncipe off the west coast of Africa. Another group went to Sobral, Brazil. When the teams returned, three members, Eddington, British Astronomer Royal Frank Watson Dyson, and Charles Davidson (one of the leaders of the Sobral party) authored the paper, "A Determination of the Deflection of Light by the Sun's Gravitational Field, from Observations made at the Total Eclipse of May 29, 1919," which was read to the Royal Society on November 6, 1919.

The paper begins with the authors' statement and the possible outcomes of their observations: "The purpose of the expeditions was to determine what effect, if any, is produced by a gravitational field on the path of a ray of light traversing it. Apart from possible surprises, there appeared to be three alternatives, which it was especially desired to discriminate between—(1) The path is uninfluenced by gravitation. (2) The energy or mass of light is subject to gravitation in the same way as ordinary matter. If the law of gravitation is strictly the Newtonian law, this leads to an apparent displacement of a star close to the sun's limb amounting to $0.87''$ outwards. (3) The course of a ray of light is in accordance with Einstein's generalized relativity theory. This leads to an apparent displacement of a star at the limb amounting to $1.75''$ outwards."

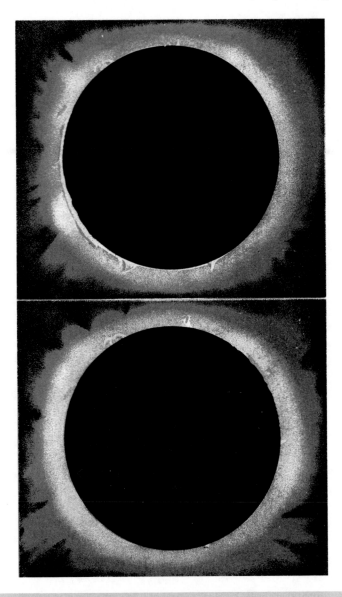

Fig. 6.12 The original caption of these images states that it shows, "Telescopic views of the real flames." Warren de la Rue sketched the corona during the total eclipse July 18, 1860. (Courtesy of M. E. Bakich library)

The authors also state why they chose the 1919 eclipse: "Attention was also drawn to the importance of observing the eclipse of May 29, 1919, as this afforded a specially favourable opportunity owing to the unusual number of bright stars in the field, such as would not occur again for many years." They, therefore, took into account the Sun's position during the eclipse (it was in the constellation Taurus the Bull) and the great length of totality, that meaning a darker sky at mid-eclipse.

The astronomers at Sobral used a 16-inch astrographic reflector, which, because of astigmatism (due to the improper figure of the mirror) showing up in test exposures, they stopped down to 8 inches. They also used a 4-inch lens, which they mounted in a 19-foot-long square wooden tube. Eddington's team in Príncipe also stopped their reflector down to 8 inches because test exposures revealed that it produced far superior star images at that aperture.

Each team captured views of totality on approximately 20 photographic plates. Two of them, called Plate W and Plate X, were the ones used to obtain the results, with two determinations being made from each plate. The careful measurements yielded results of 1.55″ and 1.67″ for Plate W, and 1.94″ and 1.44″ for Plate X, giving a mean of 1.65″. This result agrees favorably with the prediction made by the author of the theory of relativity, and the name Albert Einstein became synonymous with "genius" from that point on.

In their conclusion, the authors state, "Thus the results of the expeditions to Sobral and Príncipe can leave little doubt that a deflection of light takes place in the neighbourhood of the sun and that it is of the amount demanded by Einstein's generalized theory of relativity, as attributable to the sun's gravitational field." Wow!

A Wonderful Account

I will end this chapter with what I consider the greatest description ever of a total solar eclipse. I excerpted the following text from two sections of the book *Corona and Coronet: Being a narrative of the Amherst Eclipse Expedition to Japan, in Mr. James's schooner-yacht Coronet, to observe the Sun's total obscuration 9th August, 1896*, written by Mabel Loomis Todd and first published in 1898 by Houghton, Mifflin & Company. The first brief part comes from the introduction, and the more lengthy section is from Chapter 29.

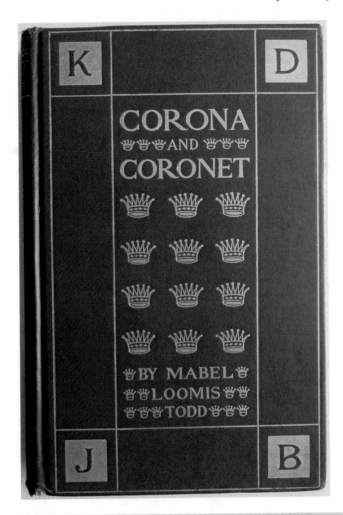

Fig. 6.13 *Corona and Coronet* by Mabel Loomis Todd. (Courtesy of M. E. Bakich library)

From the author's "Introduction":

Chasing eclipses, always of interest in itself whether the eclipse be caught or not, yields great wealth to science when these elusive phenomena are properly overtaken.

Sun and moon are of apparently the same size, and by a happy working of the celestial mechanism the moon sometimes comes directly between us and our central luminary, causing a total eclipse of the sun. But this happy state of things can by no means last longer than eight minutes. Usually in far less time sun and moon seem to slip past one another, an though for two hours or more the partial phases may continue, the duration of entire darkness is, on an average, not much over three minutes.

The astronomer wishes totality could last three hours or three months, that by the beneficent shielding of the sun's intense brightness he might have an opportunity of studying without interruption that most beautiful and mysterious sign in nature—the outflashing radiance of the corona.

From Chapter 29, "The Eclipse":

Friday, the seventh of August, dawned portentously, with a strong south wind and drifting clouds. It was very warm, and bright at intervals. By evening rain set in, and all night torrents of water fell on the roof with a noise like shot. Saturday brought more south wind, occasional rain, moving cloud. Once in a while spots of blue shone through—increasing the nerve tension. The Astronomer, cheerful, energetic, showed no sign that nature's vagaries and threats were disturbing him, but, constantly busy with final details, passed from one instrument to another, clear methodical, definite. Working of apparatus was perfect; motions were made with automatic precision, all within the time limit, all without human intervention except to press a key at the start which sent electric currents through its mysterious, ramifying nerves.

U. S. S. Olympia

THE CORONET DRESSED FOR THE FOURTH OF JULY, YOKOHAMA HARBOR

Fig. 6.14 The *Coronet*, here shown dressed for the Fourth of July, floats in Yokohama Harbor, Japan. (Courtesy of M. E. Bakich library)

Saturday toward evening the rain suddenly ceased; a fresh feeling in the wind disclosed a change to the hopeful west, bringing a superb sunset—shreds of rose and salmon and lavender glowed against a yellow background.

During the two days' rain none of our usually multitudinous callers had appeared; but by the light of sunset a dozen or more came together—guests of distinction in the town as well as the village officer and leading citizens.

Another elaborate speech was made, explaining that in the storm their hearts had failed them; they could not look at this fine apparatus, remembering our patient preparation, when a chance of cloud on Sunday might ruin everything; but that now in the light of a bright sunset they came joyfully, bringing congratulations upon the weather from the fishermen, who were said to know all signs of the sky; and with hopeful portents from a book of prophecy and a local oracle, interrogated at a neighboring shrine. This cheering oracle we believed the more readily as telegrams from Sapporo and from the Central Meteorological Observatory in Tokyo announced "Clear to-morrow!" In truth all promised happily.

Stars enough came out in the evening for final tests of the instruments, and everything was in readiness.

Directions for observing the eclipse had been written by the Astronomer, translated into Japanese, printed and distributed to inhabitants all along the pathway of anticipated darkness, and some school-teachers in the village were to ascend a fairly accessible hill nearby with implements for drawing the corona, and with a photographic instrument lent from our camp.

Sunday dawned through a heavy shower. Sunshine succeeded: cloud followed blue sky, northwest wind almost supplanted by a damp breeze from the south full of scudding vapor. And still the hours rolled on toward two o'clock and "first contact."

The Astronomer had arranged the programme of each person with exactness long before. He still kept calmly at work, giving final directions, the multitude of details resolutely kept in mind with a philosophy as imperturbable as if the skies were clear, and cloudless totality a celestial certainty. Vagaries of the western horizon, the moods of wind and prevailing drift of cirrus and cumulus had no farther power to annoy or distract. Time was too precious. It remained for the unofficial member of the party to alternate between such hope and despair that nervous prostration seemed imminent. She watched the attempt at clearing, a matter of but a few hours, and still hoped it would come in time.

At one o'clock almost half the sky was blue—two o'clock, and the moon had already bitten a small piece out of the sun's bright edge, still partly obscured by a dimly drifting mass of cloud. Half after two, and a large part of the town was ranged along the fence inclosing our apparatus, once in a while looking at the narrowing crescent, but generally at our instruments, the sober faces in curious contrast to sooty decorations from their bits of smoked glass.

And then perceptible darkness crept onward—everything grew quiet. The moon was stealing her silent way across the sun till his crescent grew thin and wan.

The Ainu suppose an eclipse is caused by the fainting or dying of the sun-god, toward whom, as he grows black in the face, they whisk drops of water from godsticks or mustache lifters as they would in the case of a fainting person.

But no one spoke.

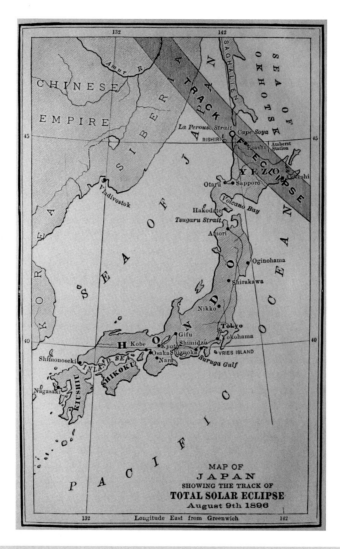

Fig. 6.15 This map shows the path of totality for the August 9, 1896, solar eclipse. (Courtesy of M. E. Bakich library)

Shortly before totality, to occur just after three, Esashi time, Chief and I went over to the little lighthouse and mounted to its summit—an ideal vantage ground for a spectacle beyond anything else it has ever been my fortune to witness.

A camera was propped up beside me, with a plate ready for exposure upon sampans and junks nearby, to test the photographic power of coronal light. Black disks, carefully prepared upon white paper, had been distributed to a number of persons, and several others were ready on the little platform, for drawing coronal streamers.

By this time the light was very cold and gray, like stormy winter twilight. The Alger rested motionless on a solid sea. A man in a scarlet blanket at work in a junk made a single spot of color.

Grayer and grayer grew the day, narrower and narrower the crescent of shining sunlight. The sea faded to leaden nothingness.

Armies of crows which had pretended entire indifference, gazing abroad upon the scene, or fighting and flapping on gables and flagpoles with unabated energy, at last succumbed and flew off in a body, friends and enemies together, in heavy haste to a dense pine forest on the mountain-side.

The Alger became invisible—sampans and junks faded together into colorlessness; but grass and verdure turned suddenly vivid yellow-green. A penetrating chill fell across the land, as if a door had been opened into a long-closed vault. It was a moment of appalling suspense; something was being waited for—the very air was portentous.

The circling sea-gulls disappeared with strange cries. One white butterfly fluttered by vaguely. Then an instantaneous darkness leaped upon the world. Unearthly night enveloped all.

With an indescribable out-flashing at the same instant the corona burst forth in mysterious radiance. But dimly seen through thin cloud, it was nevertheless beautiful beyond description, a celestial flame from some unimaginable heaven. Simultaneously the whole northwestern sky, nearly to the zenith, was flooded with lurid and startlingly brilliant orange, across which drifted clouds slightly darker, like flecks of liquid flame, or huge ejecta from some vast volcanic Hades. The west and southwest gleamed in shining lemon yellow.

Least like a sunset, it was too sombre and terrible. The pale, broken circle of coronal light still glowed on with thrilling peacefulness, while nature held her breath for another stage in this majestic spectacle.

Well might it have been a prelude to the shriveling and disappearance of the whole world—weird to horror, and beautiful to heartbreak, heaven and hell in the same sky.

Absolute silence reigned. No human being spoke. No bird twittered. Even sighing of the surf breathed into utter repose, and not a ripple stirred the leaden sea.

One human being seemed so small, so helpless, so slight a part of all this strangeness and mystery! It was as if the hand of Deity had been visibly laid upon space and worlds, to allow one momentary glimpse of the awfulness of creation.

Hours might have passed—time was annihilated; and yet when the tiniest globule of sunlight, a drop, a needle-shaft, a pinhole, reappeared, even before it had become the slenderest possible crescent, the fair corona and all color in sky and cloud withdrew, and a natural aspect of stormy twilight returned. Then the 2 minutes and a half in memory seemed but a few seconds—a breath, the briefest tale ever told.

As the beautiful corona lay there in the cloud, a soft unearthly radiance, the poetic effect as strong as if in a clear sky, the scientific value lost in vapors, it was still noticeably flattened at the poles and extended equatorially, and must have been of unusual brilliance to show so distinctly through cloud. The shape gives suggestion to astronomers as to new lines of future research.

Just after totality a telegram came from the Astronomer Royal of England, far away on the southeastern coast at Akkeshi: "Thick cloud. Nothing done."

Nature knows how to be cruel, or possibly it is mere indifference. But until, in his search after the unknown, man learns to circumvent cloud, I must still feel that she holds every advantage. On that fateful Sunday afternoon the sun, emerging from partial eclipse, set cheerfully in a clear sky; the next morning dawned cloudless and sparkling.

A few pictures of the blurred corona were taken, if of little practical use, and an interesting experiment for Roentgen rays seemed to indicate their presence in coronal light—a curious result, since they have not been found in full sunlight.

EXPEDITION MEMBERS, AND OLD SCHOOLHOUSE, AFTER THE ECLIPSE

Fig. 6.16 Members of the eclipse expedition that journeyed on the *Coronet* posed after the eclipse in front of a local schoolhouse. (Courtesy of M. E. Bakich library)

But a useful and tangible outcome of the expedition is afforded by this practical demonstration that a great number of instruments can be employed in recording the corona automatically, not only dispersing with the multitude of assistants necessary for manipulating each at critical moments, but virtually lengthening the precious minutes of totality many fold.

The corona, thus safely caught, can now be laid on our tables in manifold representations, and interrogated through the months following an eclipse until the most telling questions for its next coming are plainly evident.

And Esashi had vindicated its choice. Of all the places where meteorological observations had been made, it proved the best—clouds, that is, were thinnest.

Nothing appeared upon the plate exposed to the sampans; coronal light was not strong enough to impress them upon the sensitive surface.

But the apparatus remains—from the approach of the idea in Shirakawa, in 1887, when it was roughly but accurately carried out for the eclipse of that year; the far better evolution in West Africa in 1889 by pneumatic contrivances; and the smoothly running devices intrusted to electricity in Yezo in 1896—perfected result of three cloudy eclipses.

Chapter 7

How to Observe the Sun Safely

As we've heard a million times, observing the Sun can be dangerous without the proper precautions. While environmental exposure to ultraviolet radiation contributes to the accelerated aging of the outer layers of the eye and the development of cataracts, the concern over improper viewing of the Sun during an eclipse specifically relates to the development of "eclipse blindness" or retinal burns. This chapter details the potential risks and provides safe ways to view the Sun.

Much of this data comes directly from B. Ralph Chou, Associate Professor at the School of Optometry and Vision Science, University of Waterloo, Ontario, Canada. Dr. Chou is the recognized world expert on solar filter safety.

© Springer International Publishing Switzerland 2016   73
M.E. Bakich, *Your Guide to the 2017 Total Solar Eclipse*, The Patrick Moore
Practical Astronomy Series, DOI 10.1007/978-3-319-27632-8_7

Fig. 7.1 These are the classic Eclipse Glasses made by Rainbow Symphony. They have a cardboard housing and thin plastic lenses. Before you use them, take a few seconds to inspect them for potential problems. (Photo courtesy of the author)

As we've heard a million times, observing the Sun can be dangerous without the proper precautions. The solar radiation that reaches Earth's surface ranges from ultraviolet radiation at wavelengths longer than 290 nanometers to radio waves in the meter range. The tissues in the eye transmit a substantial part of the radiation between 380 nanometers and 1,400 nanometers to the light-sensitive retina at the back of the eye. While environmental exposure to ultraviolet radiation contributes to the accelerated aging of the outer layers of the eye and the development of cataracts, the concern over improper viewing of the Sun during an eclipse specifically relates to the development of "eclipse blindness" or retinal burns.

Exposure of the eye's retina to intense visible light causes damage to its light-sensitive rod and cone cells. The light triggers a series of complex chemical reactions within the cells that damages their ability to respond to a visual stimulus and, in extreme cases, can destroy them. The result is a loss of sight, which may be temporary or permanent depending on how severe the damage is.

When a person looks repeatedly, or for a long time, at the Sun without proper eye protection, this photochemical retinal damage may be accompanied by a thermal injury—the high level of visible and near-infrared radiation causes heating that literally cooks the exposed tissue. Oh, man, that sounds nasty! This thermal injury destroys the rods and cones, creating a small blind area. The danger to vision is significant because retinal injuries occur without you knowing it—there are no pain receptors in the retina—and the visual effects do not occur for at least several hours after the damage is done.

Fig. 7.2 Never look at the Sun before you put the glasses on. Make sure they fit closely to your eyes and that they're secure on your head, then look around to find the Sun. (Photo courtesy of the author)

The only time that you can view the Sun safely with the naked eye is during totality. **It is never safe to look at a partial or annular eclipse, or the partial phases of a total solar eclipse, without the proper equipment and techniques.** Even when 99 percent of the Sun's visible surface (the photosphere) is obscured during the partial phases of a solar eclipse, the remaining crescent Sun is still intense enough to cause a retinal burn, even though illumination levels are comparable to twilight.

Failure to use proper observing methods may result in permanent eye damage or severe visual loss. This can have important adverse effects on career choices and earning potential because most individuals who sustain eclipse-related eye injuries are children or young adults.

The same techniques for observing the Sun outside of eclipses are used to view and photograph annular solar eclipses and the partly eclipsed Sun. The safest and most inexpensive method is by projection. A pinhole or small opening is used to form an image of the Sun on a screen placed about a meter behind the opening. Multiple openings in perfboard, in a loosely woven straw hat, or even between interlaced fingers can be used to cast a pattern of solar images on a screen.

Fig. 7.3 Once you spot the Sun you can observe it for as long as you like. Be sure to look away before taking the glasses off. (Photo courtesy of the author)

You can observe a similar effect on the ground below a broad-leafed tree, as has been mentioned previously. The many "pinholes" formed by overlapping leaves creates hundreds of crescent-shaped images. Binoculars or a small telescope mounted on a tripod can also be used to project a magnified image of the Sun onto a white card.

All of these methods can be used to provide a safe view of the partial phases of an eclipse to a group of observers, but you must take care to ensure that no one looks through the device. The main advantage of the projection methods is that nobody is looking directly at the Sun. This could be especially important if children are part of the group. The disadvantage of the pinhole method is that you have to place the screen at least a meter behind the opening to get a solar image that is large enough to see easily.

The Sun can only be viewed directly when filters specially designed to protect the eyes are used. Most such filters have a thin layer of chromium alloy or aluminum deposited on their surfaces that attenuates both visible and near-infrared radiation. A safe solar filter should transmit less than 0.003 percent of visible light. This works out to a density of approximately 4.5 of visible light (from 380 to 780 nanometers) and no more than 0.5 percent (a density of approximately 2.3) of near-infrared radiation—from 780 to 1,400 nanometers.

One of the most widely available filters for safe solar viewing is shade number 14 welder's glass, which can be obtained from welding supply outlets. A popular inexpensive alternative is aluminized Mylar manufactured specifically for solar observation. Please note: So-called "Space Blankets" and aluminized Mylar used in gardening are not suitable for this purpose! Unlike a welding glass, aluminized Mylar can be cut to fit any viewing device, and doesn't break when dropped.

In the past, many experienced solar observers used one or two layers of black-and-white film that had been fully exposed to light and developed to maximum density. The metallic silver contained in the film emulsion was the protective filter. Some of the newer black and white films use dyes instead of silver, and these are unsafe. Black-and-white negatives with images on them (for example, medical X-rays) are also not safe.

Unsafe filters include all color film, black-and-white film that contains no silver, photographic negatives with images on them (X-rays and snapshots), smoked glass, sunglasses (single or multiple pairs), photographic neutral density filters, and polarizing filters. Most of these transmit high levels of invisible infrared radiation, which as I mentioned before, can cause a thermal retinal burn.

The fact that the Sun appears dim, or that you feel no discomfort when looking at the Sun through the filter, is no guarantee that your eyes are safe. Solar filters designed to thread into eyepieces that are often provided with inexpensive telescopes are also unsafe. These glass filters can crack unexpectedly from overheating when the telescope is pointed at the Sun, and retinal damage can occur faster than the observer can move the eye from the eyepiece. Avoid unnecessary risks.

Some observers have expressed concern about the possibility that the range of ultraviolet radiation in sunlight called UVA radiation, which has wavelengths between 315 nanometers and 380 nanometers, may also adversely affect the retina. While there is some experimental evidence for this, it only applies to the special case of an eye disorder called aphakia, where the natural lens of the eye has been removed because of a cataract or injury, and no ultraviolet-blocking spectacle, contact lens, or intraocular lens has been fitted in its place.

In an intact normal human eye, UVA radiation does not reach the retina because it is absorbed by the crystalline lens. In cases of aphakia, however, normal environmental exposure to solar ultraviolet radiation may indeed cause chronic retinal damage. However, the solar filter materials discussed throughout this book attenuate that radiation to a level well below the minimum permissible occupational exposure for UVA, so an observer with aphakia is at no additional risk of retinal damage when looking at the Sun through a proper solar filter.

Jay Pasachoff, professor of astronomy at Williams College in Massachusetts, made a great point regarding warnings about solar safety. He said that in the days and weeks preceding a solar eclipse, there are often news stories and announcements in the media, warning about the dangers of looking at the eclipse. Unfortunately, despite the good intentions behind these messages, they frequently contain misinformation, and may be designed to scare people from seeing the eclipse altogether.

However, this tactic may backfire, particularly when the messages are intended for students. A student who heeds warnings from teachers and other authorities not to view the eclipse because of the danger to vision, and learns later that other students did see it safely, may feel cheated out of the experience. Having now learned that the authority figure was wrong on one occasion, how is this student going to react when they receive other health-related advice about drugs, alcohol, AIDS, or smoking? Misinformation may be just as bad, if not worse, than no information at all.

As noted earlier, a #14 welder's filter is the preferred density for viewing the Sun. A couple years ago, several editors at *Astronomy* magazine posed a question to Dr. Chou asking if there were any alternatives to #14 welder's glass. He replied, "When you do the calculation of retinal exposure for a midday Sun with air mass 1, the #12 filter provides enough attenuation of solar radiation to protect the retina."

Fig. 7.5 A #14 welder's glass is a safe filter to use to observe the Sun anytime. (Photo courtesy of the author)

However, many observers have found that #12 gives an uncomfortably bright image, while #14 provides a more comfortable view. Again, this pertains to air mass 1 with the Sun at maximum altitude. As the altitude drops and air mass increases, then the shade number for comfortable vision may decrease, but how much depends on the individual. For the majority of observers, #14 does the job quite nicely, but as they say in the car ads, your mileage may vary. That's a great answer, so I recommend you use or suggest ONLY a #14 filter.

Dr. Chou went on, however, and I definitely want to repeat what he said here. "One concern I have is that polycarbonate welding filters are replacing the familiar green glass ones, and these have quite different transmittance properties in the infrared. If the filter is just green polycarbonate, then it has a fairly high infrared transmission, which is not much of a concern for welding, but could be a problem for solar observing. Polycarbonate filters with a gold coating, on the other hand, have very good infrared protection, and are comparable to the old glass filters. One has to be careful when buying polycarbonate welding filters that the gold coated one is being purchased."

He added, "It seems to me that while welding filters are a good standby, eclipse glasses with dark polymer or metal coated resin film are now so common that we should encourage people to buy those instead of the welding filters." But he also provided us the data on glass shade 12 welder's filter, which he tested on his Cary 500 spectrophotometer.

He got a luminous transmittance of 0.002 percent, which works out to a shade number of 11.9. He measured the extreme far-ultraviolet transmittance at 2.8×10^{-6} percent, the near-ultraviolet transmittance at 1.2×10^{-5} percent, and the infrared transmittance at 0.001 percent. He commented that he could see the noise level in the infrared and ultraviolet measurements because he was working at the limits of detectability of the spectrophotometer in these wavelengths and the transmittance levels are extremely low.

Chapter 8

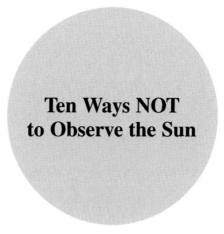

Ten Ways NOT to Observe the Sun

With the eclipse looming, people are beginning to discuss safety issues related to observing the Sun. While observing the commonsense advice in the last chapter is an important prerequisite, a lot of the warnings are going overboard.

"Don't look at the Sun!" "The Sun is dangerous!" "Never look at the Sun!" "The Sun will blind you!" If these pronouncements are all we hear between now and the time the eclipse happens, even *I* might be encouraged to stay indoors during the eclipse!

We all want everyone to view the event, and do it safely. But rather than just screaming "Don't!" and "Never!" how about if we phrase this a bit more positively: "The best way to view …" or "One safe way to view …"?

© Springer International Publishing Switzerland 2016
M.E. Bakich, *Your Guide to the 2017 Total Solar Eclipse*, The Patrick Moore Practical Astronomy Series, DOI 10.1007/978-3-319-27632-8_8

Fig. 8.1 Ron Lambert of El Paso, Texas, had his Tele Vue eyepiece capped when he accidentally moved the Sun into the telescope's field of view for less than a second. The heat caused the eyecap to melt. Imagine the damage it would do to your eye! (Photo courtesy of the author)

That said, some definite no-no's come to mind regarding solar observing, and as the eclipse gets closer, we'll need to spread the word as to what's right and what's wrong. To this end, I've built on the last chapter's advice and created a list of ten ways you never should observe the Sun. Although this chapter is short, it's really important.

Space Blankets or Telescope Covers

Some people see approved "Sun-viewing glasses" made of cardboard and a thin sheet of Mylar, and they immediately believe any similar material will serve as a safe solar filter. Not true. The material manufacturers use in the glasses is optical Mylar, a special type that filters most of the visible light and all of the truly dangerous infrared and ultraviolet radiation. Laboratories and organizations of opticians and ophthalmologists have certified this material safe. Space blankets and Mylar-infused telescope covers have totally different purposes and are not safe to view the Sun through.

Fig. 8.2 Mylar treated blankets keep telescopes from getting too hot when they're left out during the daytime. (Photo courtesy of the author)

Black-and-White Film That Uses Dyes Instead of Silver

Believe it or not, some experienced solar observers have used older black-and-white film to safely observe the Sun. This process involves completely exposing the film to light, developing it to maximum density, and stacking (taping together) at least two layers of it. Some newer films, however, use dyes instead of the silver compounds found in the older types. That makes them unsafe to view the Sun through.

Medical X-Rays or Any Film With an Image on It

This one's easy to explain. Think of the image on any photographic negative or X-ray the same way you would think of a great big hole.

Compact Discs

The silvering on compact disks, DVDs, or Blu-ray discs is not optical Mylar, and so it's not safe to view through. And even if it were safe, the plastic between the silvered layers would distort the Sun's image so badly that you'd immediately throw the disc away.

Fig. 8.3 Trying to look through a compact disk is not a safe way to view the Sun. (Photo courtesy of the author)

Smoked Glass

Smoke is ash. It also will contain other chemicals depending on what you have burned. Ash is opaque. That is, it doesn't let light through. So, if you use glass that you have smoked (probably with a candle flame), all you're doing is cutting down the amount of light and other radiation reaching your eyes. As the light dims, you might think, "Hey, it's safe to observe now." Absolutely not true. The same principle works against our eyes as the total part of a solar eclipse approaches. As 95 percent—and more—of the Sun's light disappears, people who don't know better believe it's safe to observe our daytime star. Also totally untrue. Please note that the other type of smoked glass (sometimes called frosted glass), the kind manufacturers make for effect, is also unsafe for solar viewing.

Sunglasses or Multiple Sunglasses

Just because the word "Sun" is part of this item, it doesn't mean you can look directly at the Sun through them. And "UVA" or "UVB" protection means the material will block some of the harmful solar rays, but only indirect ones, that is, those reflected off earthly objects. Do not use sunglasses—or even multiple pairs stacked—to view the Sun.

Fig.8.4 Despite their name, sunglasses are not a safe way to observe the Sun. (Courtesy of Rich Niewiroski, Jr./Wikimedia Commons)

Color Film

Like some of the newer black-and-white films, color film only contains dyes. It never has any of the silver compounds, which remain within the emulsion after you develop black-and-white film, making it safe for solar viewing. Never look through color film at the Sun, no matter how dark it appears and no matter how many layers you stack.

Neutral Density Filters

Neutral density (ND) filters reduce the amount of all wavelengths of light striking them. They come in a variety of flavors: 60 percent, 40 percent, 10 percent, even 3 percent. The numbers indicate how much of the (100 percent of) light striking the filter passes through. But what's the key word here? Yep: light. ND filters, which are great for cutting down the Moon's light, do not impede infrared or ultraviolet radiation. Oh, and here's another potentially dangerous possibility: Neutral density filters come in 1¼ and 2 sizes so they can screw into the barrels of eyepieces. But for telescopes, you should attach approved solar filter to the front of the telescope. Doing that cuts down the Sun's light before it enters the telescope. Placing a neutral density filter at the eyepiece end allows focused, unfiltered sunlight to strike it. Such a heat buildup could shatter the filter, with devastating results.

Fig. 8.5 Screwing a neutral density filter into an eyepiece and focusing the Sun's light through it could cause the filter to crack from the heat. Don't do it! (Photo courtesy of the author)

Polarizing Filters

For the most part, amateur astronomers use this accessory as a variable neutral density filter. By changing the angle of transmittance of the two filters that make up a polarizer, they can select how much light comes through. So, instead of buying four ND filters, they buy one polarizer. Unfortunately, as with ND filters, polarizers do not block infrared or ultraviolet radiation, so don't use one to look at the Sun.

Any "Solar" Filter That Screws into or Fits Over an Eyepiece

Through the years, this is the one item on this list that has had the potential to cause the most damage. Why? Because well into the 1980s, some manufacturers packaged them with their telescopes, thus putting an invisible seal of approval on them. Nobody includes them with telescopes these days. If you have one, throw it away immediately. The Sun's light, concentrated on this type of filter by a telescope's lens or mirror, has cracked many of them, sometimes with eye-damaging results.

Fig. 8.6 This device—a solar filter that screws into an eyepiece—has been distributed by telescope manufacturers through the years. Never use one! Many have cracked when focused sunlight falls on them. (Photo courtesy of the author)

Chapter 9

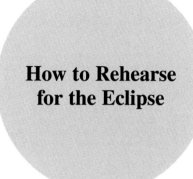

How to Rehearse for the Eclipse

As I write this, the eclipse is a long way off for most people. Some, however, are beginning to get nervous. Hey, that's what people do! They start asking questions. The more hands-on folks desire a bit more. They want to get out under the daytime sky and check out the circumstances themselves. If you're one of them, I have good news: There is a way to conduct a pretty accurate rehearsal for the eclipse.

First, some astronomy background. Our planet's axis tilts 23.5° to the plane of its orbit around the Sun. This orientation explains why we experience seasons. When the northern tip of Earth's axis points toward the Sun, it's the Northern Hemisphere's summer. When the southern tip of the axis points to the Sun, the season in the Northern Hemisphere is winter. Spring and autumn lie midway between those extremes. All seasons reverse in the Southern Hemisphere.

© Springer International Publishing Switzerland 2016
M.E. Bakich, *Your Guide to the 2017 Total Solar Eclipse*, The Patrick Moore
Practical Astronomy Series, DOI 10.1007/978-3-319-27632-8_9

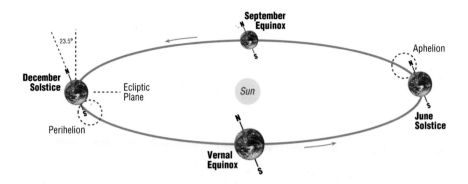

Fig. 9.1 Because Earth tilts 23½° to the plane of its orbit, we experience seasons. (Courtesy of Holley Y. Bakich)

Because of the tilt, the Sun's altitude at any location changes by 47° in the 6-month spans from June to December or from December to June. On the June solstice (the first day of summer in the Northern Hemisphere), the Sun stands as high in the sky as it will be all year. Conversely, on the December solstice (the first day of winter in the Northern Hemisphere), the Sun is as low in the sky as it gets all year.

When astronomers give the Sun's position, they use the two celestial coordinates, right ascension and declination. These two values roughly correspond to latitude and longitude on Earth. The Sun's declination varies from most southerly (December solstice) to most northerly (June solstice). Except at those extremes, then, the Sun will have the same declination twice during the year.

On August 21, 2017, the Sun's declination will be 11° 51′, approximately. In 2016, the date when the Sun's declination is closest to that value is August 21. So, if you want to "practice" observing the Sun where it will be on eclipse day, head out August 21. Maybe you want to set up a filtered telescope. Maybe you want to take a few pictures (although photographing the eclipse is not advised, especially if this will be your first). Maybe you just want to check out a prospective observing site. How far away are any trees? Buildings? Taking these factors into account, you can see how the Sun will perform on eclipse day.

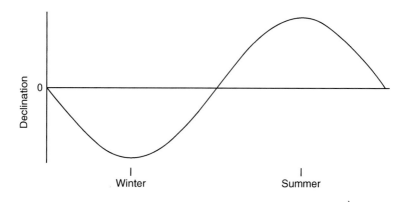

Fig. 9.2 The Sun's declination changes from 0° to 23.5° north back to 0° and then to 23.5° south throughout the year. (Courtesy of Holley Y. Bakich)

Recall that I showed there were *two* dates where the Sun had the same declination during the year. The other date in 2017 where our star's declination comes closest to 11° 51′ is April 20. On that date, the Sun's path through the sky will be the same as it will be on eclipse day in 2017. So, anything you want to try on eclipse day you can practice April 20th.

This date is the closest approximation to what you'll see on eclipse day. The Sun will rise and set around the same times, and it will cross the meridian (the imaginary north-south line that passes through the overhead point; the Sun crosses it at midday) the same time as on eclipse day. Here's something to consider, however. The Sun's declination doesn't change all that much from day to day. In fact, if your rehearsal occurs as many as 3 days before or after any of the listed date, you really won't notice any difference when August 21, 2017, rolls around.

For your convenience, here are the dates arranged into a table:

Year	First date range	Second date range
2016		August 18 to August 24
2017	April 17 to April 23	August 18, 19, and 20

Chapter 10

What Will You See Around the Sun During Totality?

I have traveled near and far to view 13 total solar eclipses. Long before each event, I carefully studied the Sun's position at mid-eclipse relative to the background stars. Something about seeing the stars during the daytime is so special that determining the stellar backdrop in advance adds a wonderful dimension to the experience.

What constellation will the Sun be in? Which bright stars are above the horizon? Where will the planets be? Stand by to find out.

© Springer International Publishing Switzerland 2016   93
M.E. Bakich, *Your Guide to the 2017 Total Solar Eclipse*, The Patrick Moore
Practical Astronomy Series, DOI 10.1007/978-3-319-27632-8_10

Fig. 10.1 What will the sky look like August 21, 2017, when the Moon totally covers the Sun? This diagram shows the eclipsed Sun at the moment of greatest eclipse and the brightest objects around it. (Courtesy of *Astronomy* magazine: Richard Talcott and Roen Kelly)

I will describe the sky above three locations spread across the U.S. as it will appear at mid-totality. For these sites I've chosen Salem, Oregon, which is pretty close to where the Moon's shadow first hits land; St. Joseph, Missouri, which is the site of the Front Page Science event; and Charleston, South Carolina, a location near the last point of land to experience the eclipse.

For each location, the positions are given of two planets, Venus and Jupiter, and two stars, Sirius and Arcturus. Sirius is the sky's brightest star. It's often called the Dog Star because of its position in the constellation Canis Major the Great Dog, one of the two hounds belonging to Orion the Hunter.

Arcturus is the luminary in the constellation Boötes the Herdsman. It's the fourth-brightest star overall, and the brightest one in the northern half of the sky. At night, you easily can find Arcturus by extending the curve of the Big Dipper's handle. Another word for "curve" is "arc," and there's an old adage that says, "Follow the arc to Arcturus."

Irrespective of location, on eclipse day the Sun and the Moon lie within the boundaries of the constellation Leo the Lion. It turns out that Leo's brightest star, Regulus, stands 1.3° east of the fully eclipsed Sun. For this event, east will be to the left of the Sun. Will you see Regulus? If your sky is cloud-free in the Sun's direction, probably, although don't expect it to be blazing forth. Some people might need optical aid such as binoculars to spot it.

Regulus is the sky's 21st-brightest nighttime star with a magnitude of 1.35. I'm certain you'll spot it through binoculars, whether or not you're looking for it. Even high-powered binoculars have enough of a field of view to take in the eclipsed Sun and the star. For example, Canon's 18×50 image-stabilized binoculars—the ones I'll be using—offer a field of view of 3.7°. And Fujinon FMTR-SX 7×50 binoculars, my wife's choice for eclipse viewing, have a 7°-wide field of view, which is pretty much standard for all 7×50 binos.

Regulus will be 1.3° to the east of the Sun. To estimate this distance, note how large the disk of the eclipsed Sun appears. That will be an easy estimate because the corona will surround it. Then look two and a half times that width to the left. This works because the rough angular size of both the Sun and the Moon is ½°.

Although the Sun is indeed moving to the east during the eclipse, it wouldn't close that kind of gap between it and the star for more than a day. The 1.3° distance to Regulus, therefore, holds true across the country. In fact, the Sun will actually be 3 arcminutes closer to Regulus during mid-eclipse at Charleston than it was at Salem. That distance works out to only one-tenth of the Sun's diameter, so you won't even notice it on images.

If you do find yourself searching fruitlessly for Regulus' position, think twice and limit your time spent on this endeavor. After just 2 or 3 seconds, I advise you to give up the hunt. It's not worth it! Spend your time looking at the corona, the overall appearance of the sky, or the brighter objects discussed shortly. Let me say it again: It's not worth spending more than a few seconds trying to find Regulus.

Now let's get to the three-city comparison. First up, mid-eclipse at Salem, Oregon. The Sun will have an altitude of 40°, and it will be in the east-southeast. Altitude is a measure of the distance an object is from the horizon. The altitude of any object on the horizon is 0°; that of an object at the zenith—the overhead point—is 90°. Midway between the two would be 45°. So, if the Sun is 40° high, that means it's nearly halfway up in the sky at Salem.

From Salem, Oregon
Sun's altitude = 40°

Object	Magnitude	Distance from the Sun	Direction from the Sun	Altitude
Mercury	3.3	10.5°	Southeast	29°
Venus	−4.0	36°	West-northwest	64°
Mars	1.8	8.3°	West-northwest	48°
Jupiter	−1.8	51°	East-southeast	−6°
Regulus	1.3	1.3°	East	39°
Arcturus	−0.04	61°	East-northeast	2°
Sirius	−1.5	57°	West-southwest	28°
Rigel	0.1°	61°	West	31°

For those of you who don't know the cardinal directions at your viewing locale, and this is understandable because you may be at an unfamiliar spot to see the eclipse, the easiest way to find them is to ask someone who appears to be in charge where north is—or to use the compass app on your phone or even an actual compass. Once you find north, south is behind you. Still facing north? Then east is on your right and west is on your left.

Venus will blaze 65° high in the south-southeast from Salem. Jupiter, however, won't be visible because at this location it hasn't risen yet. Even as far east as Casper, Wyoming, the giant planet's altitude will be a paltry 10° above the horizon at mid-eclipse. Jupiter is bright—magnitude −1.8 at this time—so you'll probably spot it from Casper, but points west of there will have issues. Binoculars will reveal the planet when it's lower in the sky, but conditions will have to be both cloud free and haze free for you to see it naked eye.

From Salem, Sirius appears 28° high in the south. That's three-quarters as high as the Sun, to our star's lower right. The star should therefore be a reasonably easy target if it's clear. Arcturus, however, stands a measly 2° above the east-northeast horizon. In other words, invisible.

Next on our list is St. Joseph, Missouri, a location on the eclipse's centerline only 200 miles east of the geographic center of the U.S. Here, the Sun will stand 62° high in the south at mid-eclipse. Venus, which you probably will have been watching for the past 20 minutes, will have an altitude of 58° in the west-southwest. Jupiter will be much easier to see from St. Joseph than from points west of there. It will be in the southeast 23° above the horizon.

From Rosecrans Memorial Airport, St. Joseph, Missouri
Sun's altitude = 62°

Object	Magnitude	Distance from the Sun	Direction from the Sun	Altitude
Mercury	3.3	10.5°	Southeast	53°
Venus	−4.0	36°	West-northwest	58°
Mars	1.8	8.3°	West-northwest	66°
Jupiter	−1.8	51°	East-southeast	23°
Regulus	1.3	1.3°	East	62°
Arcturus	−0.04	61°	East-northeast	29°
Sirius	−1.5	57°	West-southwest	19°
Rigel	0.1°	61°	West	11°

Regarding our target stars, Sirius will be 19° above the southwest horizon, so roughly one-third as high as the Sun. You shouldn't have a problem spotting it, but again, it's just one star, so don't spend a lot of time trying to pinpoint it. If you don't see it immediately with just your eyes, take a second or two to scan the area through binoculars, then move on. Arcturus will be 30° up in the east, a little higher than Jupiter. That planet, however, is five times as bright as the star, so I'm not counting on seeing Arcturus without binoculars.

Let's zoom east one more time to Columbia, South Carolina. From here at mid-totality, the Sun will be 62° high in the southwest. By coincidence, Venus and Jupiter will have identical altitudes: 41°. Sirius will stand a paltry 8° high in the west-southwest, a possibly difficult catch through the summer haze often prevalent throughout the Southeastern U.S. Arcturus, on the other hand, will be quite high: 48° up in the east. The big problem here is that, to attempt to find it, you'll have to look more than 90° to the left of where the eclipse is.

From Columbia, South Carolina
Sun's altitude = 62°

Object	Magnitude	Distance from the Sun	Direction from the Sun	Altitude
Mercury	3.3	10.5°	Southeast	58°
Venus	−4.0	36°	West-northwest	41°
Mars	1.8	8.3°	West-northwest	59°
Jupiter	−1.8	51°	East-southeast	41°
Regulus	1.3	1.3°	East	63°
Arcturus	−0.04	61°	East-northeast	48°
Sirius	−1.5	57°	West-southwest	8°
Rigel	0.1°	61°		−5°

This next tidbit is for all of you who wonder about the visibility of the other two planets. And as you can see, two of them will be much closer to the action than some of the objects listed. Mars, glowing at magnitude 1.8, lies some 8.3° west-northwest of the Sun. It's an unlikely possibility through binoculars. But an utterly impossible sighting will be magnitude 3.3 Mercury, 10.5° southeast of the Sun. Don't even try. Just be happy in the knowledge that it is, indeed, in the sky smiling down at you. Finally, the ringed planet Saturn will be invisible (below the horizon) from all locations.

Chapter 11

How Will the Sun Appear During Totality?

During totality, the Moon covers the Sun's face, so the simplest description of the Sun's appearance mid-eclipse is that our daytime star appears like a hole in the sky. As usual however, the simplest answer doesn't tell the whole story. For while the Moon does blot out the brilliant face of the Sun—what astronomers call the photosphere—it's nowhere near large enough to hide the corona, the thin outer atmosphere that becomes visible only during totality, or even prominences. This chapter, therefore, will detail what observers will see when they look toward the fully eclipsed Sun.

© Springer International Publishing Switzerland 2016
M.E. Bakich, *Your Guide to the 2017 Total Solar Eclipse*, The Patrick Moore
Practical Astronomy Series, DOI 10.1007/978-3-319-27632-8_11

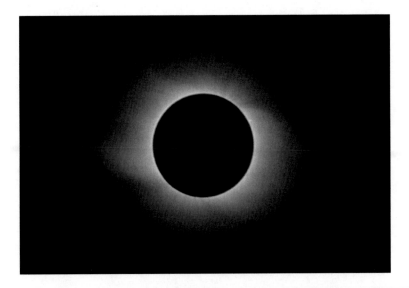

Fig. 11.1 The solar corona is the Sun's thin outer atmosphere, which only appears during totality. Our star's magnetic field dictates its shape. (Courtesy of Ben Cooper)

Here's an interesting fact: The corona glows with only one-millionth the light of the Sun's photosphere. You know what else is only one-millionth as bright as the Sun? The Full Moon! Yep, the corona—that dim, normally invisible, faint, evanescent, indistinct, barely seen, undetectable feature (here's where the thesaurus ran out of terms)—the corona actually shines as brightly as the Full Moon!

Astronomers now can predict with some certainty how the corona will appear during totality August 21, 2017. Researchers have discovered that its appearance is fundamentally different depending on whether the Sun is at or near solar maximum (the peak of its activity) or at solar minimum. The corona during an active Sun period shows many streamers at all angles around the disk of the Sun. In contrast, the corona during a quiet Sun period shows larger bottle-shaped streamers concentrated in latitudes near the equator.

Here's an example with a bit more depth—and a lot more jargon. Predictive Science Inc. issued its model prediction of the structure of the corona for the total solar eclipse that occurred November 13, 2012. Predictive Science uses a theoretical model based on magnetohydrodynamic equations to predict the state of the corona. For that eclipse the company also created an experimental version of their prediction that used radial magnetic field maps. This technique is still undergoing testing.

Obviously, the Sun's magnetic field is a key ingredient to any predictive model of the corona and how it would appear during totality. Full-disk measurements of the line-of-sight component of the photosphere's magnetic field are the most readily available measurements of the Sun's overall magnetic field.

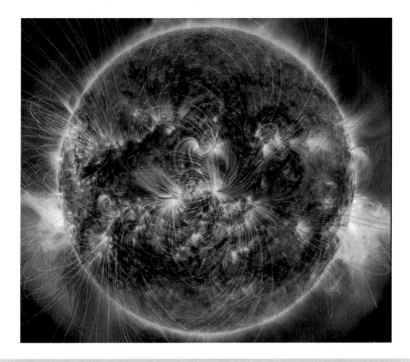

Fig. 11.2 This image of the Sun shows a potential field source surface. In other words, these are the lines of the Sun's magnetic field mapped over an image of our star. (Courtesy of NASA/SDO/AIA Science Team)

Maps of the photospheric magnetic field are produced by a number of observatories, including the Wilcox Solar Observatory at Stanford University; two facilities run by the National Solar Observatory: the Synoptic Optical Long-term Investigations of the Sun magnetograph at Kitt Peak and NSO's Global Oscillation Network Group; the 150-foot Solar Tower at Mount Wilson observatory; and the Helioseismic and Magnetic Imager aboard NASA's Solar Dynamics Observatory spacecraft. The magnetic field maps fill with data over the course of a solar rotation, so they may contain data that is as much as 27 days old.

Unfortunately, the Sun's magnetic flux is always evolving. What solar researchers ideally want is a map that captures the state of the Sun's field at a given moment. Obviously, for solar eclipses that moment is sometime during totality. Flux evolution models have been successful in reproducing many of the observed properties of the photosphere's fields, and they also can estimate the likely state of its magnetic field on unobserved portions of the Sun.

The newest model, based on a flux evolution model introduced in 2000, accounts for the known transport processes in the solar photosphere—differential rotation, meridional flow, supergranular diffusion, and random flux emergence.

The new model improves on the older one by incorporating rigorous data assimilation methods into it, especially those from the National Solar Observatory. It's therefore safe to say that we'll have a pretty solid idea as to the shape of the Sun's corona well prior to August 21, 2017. After the event, astronomers will compare all the predictions to the actual corona and refine the theory a bit more. The prediction for future eclipses will only get better and better.

During totality the corona will be the thing to watch. But there are two other super-cool features you can spot ever-so-briefly at second and third contacts. Astronomers call one of them Baily's beads because French astronomer Francis Baily provided the first explanation for them way back in 1836. And now French eclipse-chaser Xavier M. Jubier has developed an application called Solar Eclipse Maestro, which, among other features, is able to model how Baily's beads will appear from any location. The software uses a high-resolution plot of the lunar limb profile and displays an animated prediction for how the beads will appear as things move from second to third contact.

Fig. 11.3 Baily's beads form as sunlight streams through gaps created by mountains and valleys at the edge of the Moon. (Courtesy of Mike Reynolds)

When I asked Xavier if he could generate a prediction specifically for Rosecrans Memorial Airport, he had it to me within a day! And not only the whole Sun but also close-ups at both second and third contacts. You can find a link to the animations he created on the www.stjosepheclipse.com website on the "Education" page, or directly at www.fpsci.com/baily's.html. Merci, Xavier.

The second feature you'll see at second and third contacts has a name that—when I hear it shouted at total solar eclipses—makes for the two most exciting words in science: Diamond Ring! Think of each diamond ring as either the last of Baily's beads you'll see at second contact or the first one you'll spot at third contact.

Apart from the corona and Baily's beads, there's one other feature we may be lucky enough to see during totality—prominences. A prominence is a localized eruption of bright material from the Sun's surface sculpted by our star's magnetic field. Prominences start at the photosphere and extend into the corona. The shortest prominences are a few thousand miles long, and the very biggest can reach a half million miles in length. There's one other part in the definition of a prominence: It occurs along the Sun's edge so we view it against the black background of space from here on Earth. A prominence that we happen to see silhouetted against the Sun's disk is called a solar filament. It appears dark because the gas in prominences is a bit cooler than our star's surface.

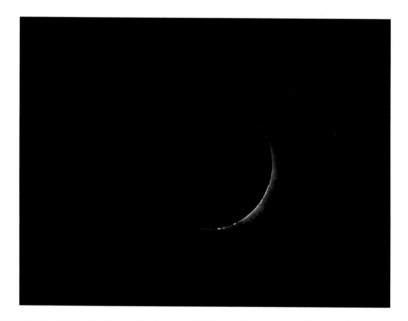

Fig. 11.4 The reddish "bumps" at the solar limb are prominences, eruptions of gas that glow against the blackness of space. (Courtesy of Mike Reynolds)

Totality is not a requisite to see prominences. Any telescope equipped with a Hydrogen-alpha (or, H-alpha) filter will reveal them. That's because the plasma in prominences is the same as the stuff that's in the Sun's chromosphere, the layer just above the photosphere. This extremely thin material emits H-alpha light, which appears quite red. Normally, the brilliance of the photosphere swamps H-alpha light. An H-alpha filter, however, blocks all the wavelengths *except* the one for which it's named, so it enables observers to see both the chromosphere and prominences easily.

Chapter 12

Earthly Effects to Look for During the Eclipse

Cool things are afoot before and after totality. Although the big payoff is the exact lineup of the Sun, the Moon, and each reader's location, you should keep your eyes open during the partial phases that lead up to the eclipse and especially those that follow it. The reason I emphasize the latter half of the eclipse is that before totality, people (yes, me included) are usually so amped up that it's hard to concentrate on what's going on. This chapter will outline everything to watch between first and second contacts and between third and fourth contacts.

As you look toward the Sun through a safe solar filter to view the beginning of the eclipse, the universe will set your mind at ease when you see the Moon take the first notch out of the Sun's disk. That moment is known as first contact. People have told me that first contact isn't all that big of a deal, but when I see it my blood starts racing because I know what comes next. If you're part of a group, especially if someone is observing through a telescope, you really don't have to worry about the exact timing of the start. I guarantee that someone—probably a scope-user—will yell, "First Contact!"

© Springer International Publishing Switzerland 2016   105
M.E. Bakich, *Your Guide to the 2017 Total Solar Eclipse*, The Patrick Moore
Practical Astronomy Series, DOI 10.1007/978-3-319-27632-8_12

Fig. 12.1 First contact occurs when the Moon's disk touches the Sun and the eclipse begins. In this great sequence, it's the second image of the Sun from the top. (Courtesy of Ben Cooper/ LaunchPhotography.com)

Then, the wait begins for second contact, the point at which the total phase of the eclipse starts. The length of time between first contact and second contact varies depending where you are on the eclipse path. If you're on the centerline at the Oregon coast, hoping to be the first person on land to see the event, you'll wait 71 minutes and 25 seconds between first and second contacts. For the Front Page Science group at Rosecrans Airport in St. Joseph, Missouri, that duration is 85 minutes and 45 seconds. And if you're near the point of greatest eclipse, just over 88 minutes will elapse between the start of the eclipse and the start of totality.

During this interval, you'll feel the general excitement level climb, but slowly. If you don't have a safe solar filter (and, really, why wouldn't you?), you won't even notice any change over the next hour. That's right. Even when the Moon covers two-thirds of the Sun, the part that remains visible is so bright that your site will still look like a sunny day. Those with filters, however, will be stealing glances at the Sun about every 30 seconds or so. They'll be able to note the slow progression of the Moon's black disk across the Sun's brilliant face.

Around the three-quarters mark, which will be roughly 20 minutes before totality, you'll start to notice that shadows are getting sharper. The reason is that the lit part of the Sun's disk is shrinking, literally approaching a point, and a smaller light source produces better-defined shadows. Here is where trees may see their leaves act like pinhole cameras as hundreds of crescent Suns appear in the gaps intermingled within their shadows. Be sure to check your own shadows occasionally. A few minutes before totality, they will be razor-sharp.

Fig. 12.2 The gaps between leaves of a tree will act as pinhole cameras. In this picture, hundreds of crescent Suns appear on the side of the author's shop during the June 10, 2002 annular solar eclipse, which was a 63 percent partial eclipse from El Paso, Texas. (Photo courtesy of the author)

At about 85 percent coverage, either you or someone you're with will see Venus 34° west-northwest of the Sun. Venus pops out first because, next to the Sun and the Moon, it's the brightest object we ever see in the sky. If it's an early sighting, or if the person first saw Venus through binoculars, the sky will still be pretty bright, so you'll probably also need binoculars to locate it. Once you find it with your naked eyes, you'll see it consistently until long after totality.

Jupiter may be a different story. It lies 51° east-southeast of the Sun and blazes at magnitude −1.8. Rather than trying to figure out that distance and direction from the Sun, note that, from St. Joseph, Jupiter will stand 24° above the east-southeastern horizon at mid-eclipse. Start looking for the giant planet about 10 minutes or so before totality. Don't spend an irrational amount of time hunting for it, though, because there's a lot more to see and Jupiter surely will pop out during totality. You can take a few seconds to glance at it then.

While the sky changes, here on Earth nature will take heed. Depending on your surroundings, as totality nears you may experience things quite strange to you. Look. You'll notice a resemblance to how the sky appears as night approaches. It's not exactly the same and, quite frankly, it's a bit hard to describe. As you look in the Sun's general direction, you'll notice that part of the sky is darker than the areas that lie around the horizon.

Next, listen. Usually as totality approaches and the atmosphere above your observing site cools, any breeze will dissipate and birds (many of whom will come in to roost) will stop chirping. It is quiet. I like to describe it as Nature holding her breath.

Finally, feel. Because the eclipse occurs in August, many parts of the country will still be pretty hot. As the sunlight striking your skin decreases, you'll feel more comfortable. A 10–15 °F drop in temperature is not unusual. How much it falls at your location depends on lots of factors, including barometric pressure and humidity.

Seconds before totality, you may have the opportunity to watch for the Moon's shadow.

If your viewing location is at a high elevation, or even at the top of a good-sized hill, you may even see the Moon's shadow approaching. This sighting is rare because of the speed of the advancing darkness. As the shadow crosses Rosecrans Airport in St. Joseph, for example, it is moving at 1,584 mph, or twice the speed of sound. Another way to spot the shadow is as it covers thin cirrus clouds if any are above your site. Again, you'll be surprised how fast the shadow moves. This said, I don't know a single eclipse-watcher who would wish clouds above their location just to glimpse the Moon's approaching shadow.

About two seconds before totality starts, the diamond ring will come into view. This sight is the combination of two features: the Sun's corona (its usually invisible outer atmosphere) and the last minuscule piece of its brilliant disk. Wait till you see it. It really does look like a diamond ring.

Besides what you see going on around the Sun during totality, take just a few seconds to tear your eyes away from the sky and scan the horizon. You'll see sunset colors all around you because, in effect, those locations are where sunset (or sunrise) are happening. Eclipse-o-philes call this the 360° sunset. And let me say one more thing about sky lighting. Because the Moon's shadow is traveling from the northwest to the southeast, the northwestern horizon will appear darker at the start of totality. That's because only the front edge of the shadow has covered you, and more is to come. Likewise, the southeastern horizon will appear darker at the end of totality because most of the shadow already has passed over your location.

Fig. 12.3 During totality, sunrise/sunset colors will appear around your entire horizon. (Courtesy of Mike Reynolds)

People usually are excited to talk about what you can see between second contact and third contact, which astronomers define as the moment totality ends. Let me add a few words about the time between third contact and fourth contact — the moment the Moon's disk no longer covers any of the Sun's face.

Here are a few positive words: Success. Gratitude. Awe. And now a few words that aren't as positive: Longing. Emptiness. Some people who will see this eclipse will have been waiting eagerly for years. And then, after a maximum — a MAXIMUM — of 161 seconds, the most spectacular sight in nature — totality — is over. And there's no chance of seeing it again until July 2, 2019, the date of the next total solar eclipse. You will have to travel for that one, though. The maximum duration of totality — a worthy 4 minutes and 32 seconds — occurs ridiculously far out in the Pacific Ocean, about a third of the way from Peru to New Zealand, some 1,300 miles northeast of Pitcairn Island, usually described as the most remote

Fig. 12.4 A souvenir is a good way to remember the event. The author purchased the pins shown at two total solar eclipses and the two most recent transits of Venus. (Photo courtesy of the author)

inhabited region on Earth. If you want to be on land, the best site will be along Highway 5 roughly 25 miles north of La Serena, Chile. Totality there will last 2 minutes and 36 seconds.

There will be plenty of time to figure out where you'll see your next total solar eclipse. But this one's not even over yet. You can watch as the Moon continues across the Sun's face for more than another hour. Venus is still visible. Jupiter, too. For how long? That depends on how clear it is above your site and how well you can pinpoint their locations before returning sunlight starts to overwhelm them. Seem anticlimactic? It is.

This also is the time to get social. You'll find yourself chatting up what you just experienced with family, friends, and even total strangers. You all shared something special, but each person's take on it will be unique. Now's the time to cement what you saw in your memory. It's also the best time to take pictures. You can ask groups of people to gather without worrying about interrupting their viewing preparations. And don't forget to take some pix of the equipment people set up and of the site itself. Can you say "Facebook"?

If you've been to a large enough event, you also might have the opportunity to buy a souvenir or two, perhaps a T-shirt or button. And remember to hold onto your eclipse glasses. You can safely view the Sun through them anytime.

By the way, there's one really good reason you might want to linger at the site until fourth contact, and it may trump all the others: traffic. Some people, especially those with small children, will begin bailing just a few minutes after totality ends. You'll be best served to let the most anxious drivers get out of your way first.

Chapter 13

Pick the Right Binoculars for the Eclipse

An Introduction to Binoculars

Some amateur astronomers consider binoculars an optional accessory, but most regard them as a necessity. Indeed, binoculars are the first thing that a beginning amateur should purchase to view the sky. If your interest in astronomy continues, then upgrade to a telescope. Most of the advanced observers I know own several binoculars. And nothing is better for enhancing your view of totality.

However, do not view *any* of the solar eclipse's partial phases through unfiltered binoculars. Only totality, when the Moon completely covers the Sun's disk, is safe to view with your unprotected eyes or through binoculars. That said, some manufacturers make approved solar filters that will fit over the front lenses of your binoculars (see Chapter 16). With those in place, you can view the Sun at any time.

© Springer International Publishing Switzerland 2016   113
M.E. Bakich, *Your Guide to the 2017 Total Solar Eclipse*, The Patrick Moore
Practical Astronomy Series, DOI 10.1007/978-3-319-27632-8_13

Fig. 13.1 Never use unfiltered binoculars to look at the Sun (except during the total phase of a solar eclipse), even if you are wearing approved solar glasses. The filters in the glasses were not designed to block that much incoming light. Also, the intense light and heat focused on the Mylar in the glasses could melt them with disastrous results. (Photo courtesy of the author)

In some ways, binoculars may be a better choice than a telescope, especially if you're just starting out in astronomy. Binoculars have a wide field of view and provide right-side-up images, making objects easy to find. They require no effort or expertise to set up—just sling them around your neck, step outside, and you're ready to go. That portability also makes binoculars ideal for those clear nights in the middle of the week when you don't have the time or inclination to set up a telescope. And, for most people, observing with two eyes open rather than one seems more natural and comfortable.

Also, binoculars are easier on your wallet. Unless you're considering image-stabilized models, binoculars offer a more affordable way to tour the sky than a telescope. If you're a parent hoping to foster a child's interest in the universe, binoculars are a good first step. Even if the appeal of stargazing eventually wanes, binoculars can be used for more down-to-Earth pursuits.

In common usage, you may hear someone refer to binoculars as a "pair of binoculars." Not exactly. The unit is simply binoculars. Technically speaking, you could refer to them as a pair of monoculars, but you'll never actually hear that.

The Numbers

Every binocular has a two-number designation, such as 7×50. The first number (in this case, 7) is the magnification, or power. The second number (50) is the diameter in millimeters of each of the objective lenses.

Fig. 13.2 Celestron's Oceana 7×50 binoculars offer a great combination of magnification and light-collecting ability. (Photo courtesy of Celestron)

This example, 7×50, is the binocular I recommend as a first unit for astronomy. A magnification of 7 is in the "medium" range, just high enough to bring out some detail in large astronomical objects. Too high a magnification will over-magnify involuntary motions of your hands. That motion causes celestial objects to move around. A high magnification also limits the field of view, making objects more difficult to find for beginners.

That being said, most advanced amateur astronomers can hand-hold binoculars up to 16× for short intervals. Also, binoculars with higher power (like a telescope with a shorter focal length eyepiece) will reveal fainter stars by increasing the contrast between the star images and the sky background. Here's a tip: When hand-holding high-power binoculars, position your hands toward the front of the binoculars, and the view will be steadier.

In our example, 50 millimeters is the aperture or size of each of the front lenses. The larger this number, the more light the binoculars collect, and that makes the target brighter. Binoculars with 50-millimeter front lenses collect more than twice as much light as those measuring 35 millimeters across. Astronomically speaking, this is a gain of nearly 0.8 of a magnitude in light-grasp. The disadvantages to larger front lenses are that the binoculars will be larger, heavier, and more expensive.

Optics

The components inside binoculars that bend the light so the image appears right-side up are the prisms. Binocular prisms come in two basic designs: roof prisms and Porro prisms (capitalized because they're named for 19th-century Italian inventor, Ignazio Porro). Roof prisms are more lightweight and smaller, but I don't recommended them for astronomy.

Fig. 13.3 Binoculars use prisms to bend incoming light several times before it reaches your eyes. Manufacturers have to position all optical components correctly so that no light is lost. (Courtesy of Holley Y. Bakich)

Porro prisms are better and are made of either BK-7 or BaK-4 glass. BaK-4 prisms (barium crown glass) are the highest quality available. BK-7 prisms (borosilicate glass) are also good quality, but image sharpness falls off slightly at the edge of the field of view compared to the view through prisms made of BaK-4.

Most high-quality binoculars are multicoated on all optical surfaces. You'll see this referred to as "fully multicoated." In the 1950s, manufacturers developed coatings (most notably magnesium fluoride, MgF) that reduced light loss and internal reflections.

Mechanical Considerations

Center-focus binoculars move both optical tubes simultaneously. Other models let you focus each tube independently. All else being equal, choose binoculars with individual focus. Center-focus units add a bit of mechanical complication. Even center-focus binoculars must allow you to adjust one of the tubes, because most people eyes do not come to the same focus.

Individual-focus binoculars tend to be more rugged and weatherproof. In either case, once you focus your binoculars on a sky object, the focus will be good for all other objects because all appear infinitely distant through binoculars.

Exit pupil

One of the most important terms when dealing with binoculars is *exit pupil*. This is the diameter of the shaft of light coming from each side of the binoculars to your eyes. If you point the front of the binoculars at a bright surface, light, or the sky, you'll see two small disks of light exiting at the eyepieces. These images show the lenses.

Fig. 13.4 The illuminated openings of these binoculars are the exit pupils. (Photo courtesy of the author)

The diameter of the exit pupil equals the aperture divided by the magnification. So for our 7×50 binoculars, the exit pupil diameter would be 50 divided by 7, or roughly 7 millimeters. For astronomy, you want to maximize this number because the pupils in our eyes dilate in darkness. The wider the shaft of light, the brighter the image will be because light is hitting more of our eye's retina.

This loose rule, however, is only true up to a point. If a binoculars' exit pupil is too large to fit into your eye, you will lose some of the instrument's incoming light.

A few rare individuals have dark-adapted pupils measuring nearly 9 millimeters in diameter. Others have small ones less than 5 millimeters across. Our pupils are largest when we're young. From age 30 on, they start to contract, but the process slows in our later decades. Women of the same age tend to have larger pupils than men, on average. Unfortunately, there's no hard and fast rule that correlates pupil size with age.

You can measure your pupil size with a gauge available from some telescope suppliers. You also may be able to get one from a pharmaceutical company or from your optometrist.

Field of View

One number you'll usually find printed directly on the binoculars is *field of view*, often abbreviated "FOV." High-quality binoculars (especially those designed for astronomy) give this number in degrees, and other models often state XXX feet at 1,000 yards. The latter lists the width of the field of view (in feet, not yards) at a distance of 1,000 yards.

If your binoculars are the latter type, or if they contain no information on the field of view at all, don't worry. There's an easy way to figure out the approximate angular diameter (in degrees) of the field of view. Just start with the same math that you used to figure out exit pupil: Divide the aperture (in millimeters) by the power. So, 7×50 binoculars will have a field of view of 50/7 = 7.14°, or, what you'll more usually see and hear, 7°.

Eye Relief

Eye relief is a function of the eyepiece. It's the manufacturer's recommended distance (for best performance) your eye's pupil should be from the eyepiece's exit lens. Eye relief generally decreases as magnification increases. Eye relief less than 10 millimeters requires you to position your eye quite close to the eyepiece.

Short eye relief poses no problem for advanced amateurs, but for beginners, longer eye relief allows the head more freedom of movement. Also, those who need to or choose to wear eyeglasses will require longer eye relief. The difference between "needing" and "choosing" depends on the type of eye problem your glasses correct for. If you are nearsighted or farsighted, you don't need glasses when you use binoculars or a telescope. The unit's focuser will correct for both of those conditions. If, however, your glasses correct for astigmatism, you will *need* to keep them on when you use optics.

If you're given a choice, select binoculars that have an eye relief between 15 and 20 millimeters. If the eye relief is not listed, and if you wear glasses, be sure to try the binoculars out *with your glasses on* before you buy them.

A Buyer's Guide

How should you choose your binoculars for the eclipse? The best advice for many amateur astronomers is to conduct a short test before their purchase.

First, pick up the binoculars and shake them gently. Then twist them gently. Then move the focusing mechanisms several times. Then move the barrels together, then apart. What you're assessing is quality of workmanship. If you hear loose parts or if there's any play when you twist or move the binoculars, don't buy them. Another thing to consider at this point is the weight of the binoculars. If you're going to be hand-holding them, try to imagine what they will feel like at the end of a long observing session.

Fig. 13.5 If at all possible, handle the binoculars before you buy them. This will help you assess their workmanship and also let you know how much they weigh. (Courtesy of Celestron)

Look into the front of the binoculars and check for dirt or other contaminants. Ignore if there is a small amount of dust on the outside of the lenses, but the inside of the binoculars should be immaculate. If not, don't buy them.

Hold the binoculars in front of you with the eyepieces toward you. Point them at a bright area. You'll see the exit pupils—disks of light formed by the eyepieces. They should be round. If they are not round, the optical alignment of the binoculars is bad and the prisms are not imaging all the light.

Crucially, you must look through the binoculars as well. Try to do this outdoors and at night. Nothing will reveal flaws in the design of binoculars more than star images. If it's impossible to test the unit at night, or even outdoors, look through a door or window at distant objects. How well do the binoculars focus? Are objects clear? If there is any sign of a double image, the two barrels are not aligned. Don't buy them.

If you are wearing glasses—and if you plan to observe with them on—there are other questions. Can you get your eyes close enough to the binoculars to see the entire field of view? Aim the binoculars at a straight line such as a phone wire or the horizon, if possible. Does the line look distorted? A tiny amount of distortion near the edges of the field of view is not a big problem. But if you see a high level of distortion, steer clear.

Repeat the above tests with several different binoculars. Once you become more familiar with how binoculars compare, you will be well on your way to purchasing an excellent unit.

One factor you may hear about is *field curvature*. This optical flaw is present in all binoculars to some extent. Field curvature results from the lens forming a sharp image on a curved surface. When the eyepieces are set to meet the part of the image that is in focus, say the center, you must change the focus to make the edges of the image sharp.

The quality of the binoculars is directly proportional to the extent that the unit minimizes this problem. An excellent instrument shows field curvature only at the edge of the field. When the entire field of the binoculars is in focus, manufacturers call it a flat field. If you use a binocular that has a flat field you won't soon forget it. Such units are expensive, however, because a binocular that is flat across most of the field requires top-notch optics.

Image-Stabilized Binoculars

What happens to an image when binoculars apply too high a magnification has already been discussed. They over-magnify the motion of your hands, creating less-than-desirable views. But what if someone could find a solution to this problem? After all, much of the allure of the night sky to amateur astronomers comes by studying it under ever-increasing magnification.

Fig. 13.6 Image-stabilized binoculars, like this 18×50 model made by Canon, reduce the effect of shaky hands. (Photo courtesy of the author)

To enjoy both a wide field of view—the kind binoculars provide—and moderately high magnification in hand-held binoculars is to have the best of both worlds. Manufacturers have achieved this by creating image-stabilized (IS) binoculars.

IS binoculars use different methods to stabilize the image. Some have batteries that power a gyroscopic mechanism. Others use a non-powered design, which relies on a gimbaled prism. In all designs, you push a button to engage the stabilization. The results are dramatic. For example, I'd never been able to hand-hold binoculars steady enough to obtain good, long looks at Jupiter's moons until I used IS binoculars.

The optics and mechanics of IS binoculars vary in the same way as the optics of regular binoculars, so you should apply the same tests I listed above if you are considering purchasing a unit. And remember, technology like this isn't free, or cheap. IS binoculars large enough to interest amateur astronomers cost, at this writing, between $1,000 and $2,000.

Giant Binoculars

Giant binoculars have front-lens apertures greater than or equal to about 4 inches (102 millimeters). Classifying giant binoculars by size is a bit of a gray area, however. Some manufacturers label the largest binoculars they sell "giant," which may be true for them. When most amateur astronomers imagine giant binoculars, however, they think of two short-focal length 4-inch refractors connected together—or something even larger!

Fig. 13.7 Celestron's Skymaster 25 × 100 binoculars fall into the "giant" category. Each barrel, essentially, is a 4-inch telescope. (Courtesy of Celestron)

As with smaller models, magnification varies. Some giant binoculars allow you to change eyepieces to increase or decrease the magnification. No giant binoculars can be hand-held. They are simply too heavy to hold. If a case didn't come with your giant binoculars, buy one. Don't try to save money at this point because you'll regret the decision later.

Maintenance

Caring for your binoculars is easy. Most units come with lens caps, eyepiece caps, and a case. Use them. These will help protect your binoculars from dust and moisture. Don't leave your binoculars exposed to direct sunlight, even if they're in their case. Most binoculars (and their cases) are black or darkly colored and will absorb a lot of heat. Heat will cause the carefully placed elements of binoculars to expand and later contract—not a good scenario.

Cleaning is only sensitive when it involves the lenses. If your lenses become dusty, blow them off with compressed air or brush them with an approved optics brush. You'll find both of these at any camera store. If you must wipe the lenses, use only lens paper, and change it frequently, rather than using the same piece to wipe back and forth.

The body of your binoculars also will get dirty. When it does, simply wipe it with a damp cloth. And because binoculars are aligned optical equipment, keep the vibrations, especially impacts, to a minimum.

Three things to know about binoculars

- Binoculars give a right-side up image.
- Binoculars let you use both eyes to observe.
- Binoculars give the best views of the biggest sky objects.

Binocular Mounts

Image-stabilized binoculars are a tremendous advance for amateur astronomy. But for the steadiest images possible, nothing beats mounting your binoculars to a tripod or custom binocular mount. Smaller, well-mounted binoculars with less magnification will, after only a few minutes of continuous use, beat hand-held binoculars of larger aperture and magnification.

Fig. 13.8 A basic binocular mount is a small accessory that lets you connect your binoculars to a photographic tripod. (Courtesy of Celestron)

The simplest binocular mount is a metal "L" bracket. Attach it to the 1/4-20 mounting hole on the binoculars' center post. The other end of the L attaches to a camera tripod. This setup is generally adequate if the objects you're observing aren't too high in the sky. For objects near the zenith (the overhead point), tripod-mounted binoculars are uncomfortable—and in some cases impossible—to use.

Another option is to purchase or build a custom-made binocular mount. Plans for binocular mounts are readily available. If you are mechanically inclined, you can build your own.

Fig. 13.9 The author built this parallelogram-style binocular mount. Because it has a movable counterweight, it can hold binoculars of all sizes and weights. (Photo courtesy of the author)

Most amateur astronomers purchase commercially made mounts. Such units employ a design based on a movable parallelogram. This arrangement keeps the binoculars pointed at an object over a wide range of motion, allowing people of varying height to use them. This setup is ideal for observing sessions or star parties where both adults and children will be viewing the same objects over the course of the session.

When selecting a binocular mount, choose one that is sturdier than you require. That way, you can easily upgrade your binoculars to a larger and heavier model in the future. A binocular mount is rugged if, a few seconds after you have located an object, the image settles down and shows no vibration (unless a strong wind is blowing). If the image is not stable, your binocular mount may be at fault. Or, the fault could lie with the second piece of equipment you'll need.

The other necessity for high-power binocular observing is the tripod the mount attaches to. Most camera tripods are inadequate for this purpose. They simply are not robust enough to handle the weight of the binoculars plus the weight of the mount, and balance also could be a problem. If you have a tripod, by all means try it. You'll know immediately if it's up to the task.

A tripod can fail in more ways than by being flimsy. It may be that your sturdy tripod, even at full extension, is not high enough to allow you to stand under your binoculars and view objects near the zenith.

Suggestions for the Eclipse

Now that you know what goes into buying binoculars, which one should you choose? First, I can recommend six manufacturers: Canon (www.usa.canon.com), Celestron (www.celestron.com), Fujinon (www.fujifilmusa.com), Nikon (www. nikonusa.com), Orion (www.telescope.com), and Vixen (www.vixenoptics.com). These companies make high-quality products, they don't make wild claims, and they stand behind their products.

Most consumers reading this will choose Celestron, Orion, or Vixen products because those manufacturers offer lower priced though still high-quality models. The other three companies produce "high-end binoculars, and Canon makes only image-stabilized models.

The main consideration then, is choosing the numbers. That's right, the two numbers I explained at the beginning of this chapter. Consider the following examples.

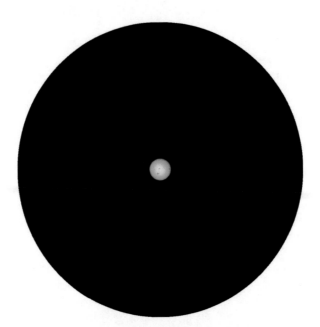

Fig. 13.10 The field of view through 7×50 binoculars is 7° wide. When you observe the eclipse through such binoculars, the Sun's disk (which has an average angular size of approximately 0.53°) will be 7.6 percent as wide as the field of view. (Image courtesy of SOHO/NASA; graphics courtesy of Holley Y. Bakich)

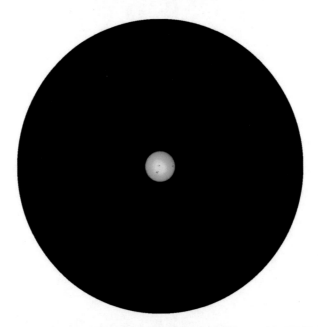

Fig. 13.11 The field of view through 10×50 binoculars is 5° wide. When you observe the eclipse through such binoculars, the Sun's disk will appear 10.6 percent as wide as the field of view. (Image: SOHO/NASA; graphics: Holley Y. Bakich)

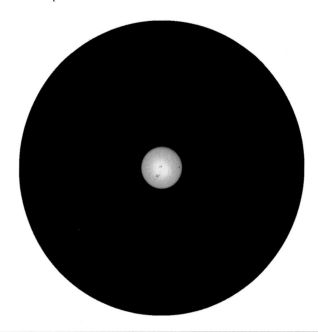

Figure 13.12 The field of view through Canon's 18×50 binoculars is 3.7°. (NOTE: Ordinarily, the field of view through such optics would be 2.8° wide.) When you observe the eclipse through such binoculars, the Sun's disk will be 14.3 percent as wide as the field of view. (Image courtesy of SOHO/NASA; graphics courtesy of Holley Y. Bakich)

There are two suggested ways for determining exactly what the Sun will look like through your binoculars. (1) Observe the Full Moon. The disks of the Sun and the Moon are the same size. (2) Purchase approved solar filters for your binoculars (see Chapter 16), and begin observing the Sun now.

That said, so far I've talked about only how the Sun's *disk* will appear through your binoculars. Remember, however, that during totality you also will see the corona, our star's thin outer atmosphere. During some eclipses it has stretched nearly three times the disk's diameter from the Sun in all directions, which would make it 3.5° across at maximum.

For most of the 12 eclipses I have observed, I've used 7×50 binoculars. But during the past three or four, I've gone with 18×50 image-stabilized models. Those definitely have a smaller field of view (3.7°), but they make the streamers in the corona a bit easier to see.

One reason to go with binoculars that have a wider field of view is the hope of capturing a planet or star and the eclipsed Sun in your field of view. Remember that on August 21, 2017, a bright star will lie near the Sun. Regulus (Alpha [α] Leonis), which shines at magnitude 1.3, will be 1.3° east of our daytime star at mid-eclipse. The next-nearest even reasonably bright object is Mars, at magnitude 1.8, which stands 8.3° to the west-northwest—indeed, too far away to capture with the Sun in a single view even through the widest of our example binoculars.

Other viewing targets aside, you'll see the main spectacle best using just your naked eyes. I will have binoculars around my neck for the eclipse. If I had to guess, I'd estimate that I'll view the roughly 2 minutes and 40 seconds of totality like this: 2 minutes with naked eyes; 40 seconds through binoculars. That seems right. But as they say, your mileage may vary.

Chapter 14

Pick the Right Telescope for the Eclipse

Buying a telescope, like buying a car, is subject to your tastes as a consumer. In other words, the choice is up to you—but I want to provide some guidelines to make your purchase as informed as possible. This chapter describes the three main types of telescopes along with a bit of history and some general advantages and disadvantages for each. I also discuss mounts and tripods, eyepiece, and accessories. Buying a telescope is a big decision, and you shouldn't rush into it just because the total solar eclipse is coming. I offer this chapter to get you started.

What do all the numbers mean?

The main number of any telescope is its aperture, or size of the main lens or mirror. Manufacturers give it in inches. Numbers like "f/4," "f/10," or "f/15" are a telescope's focal ratio. This number can give you the approximate length of any telescope. For example, a 6-inch f/4 telescope will be approximately 6 inches times 4, or 24 inches long (excluding lens shades, etc.). A 6-inch f/10 telescope, on the other hand, will be roughly 60 inches long.

Refracting Telescopes

Refraction is the bending of light that happens when it passes from air to glass and back. A refracting telescope or refractor uses this property via a carefully made lens system. Because the surfaces of the lenses have the proper shape, the light comes to a focus.

© Springer International Publishing Switzerland 2016

M.E. Bakich, *Your Guide to the 2017 Total Solar Eclipse*, The Patrick Moore Practical Astronomy Series, DOI 10.1007/978-3-319-27632-8_14

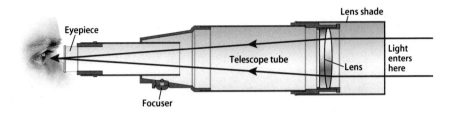

Fig. 14.1 A refractor uses a lens (a combination of two to four glass pieces) to bring light to a focus. (Courtesy of *Astronomy* magazine: Roen Kelly after Celestron)

Fig. 14.2 Celestron's PowerSeeker 60AZ is an example of a small, low-priced refractor. It has a 2.4-inch lens, sits on a stable mount, and produces right-side-up images with the supplied diagonal. (Courtesy of Celestron)

Dutch spectacle-maker Hans Lipperhey constructed the first telescope in 1608. His patent application described "an instrument for seeing faraway things as though nearby." The tube magnified objects approximately three times. Italian inventor Galileo Galilei then built his own telescopes beginning in 1609. Galileo was the first to use the new device to study celestial objects, and what he saw revolutionized astronomy forever.

The earliest telescopes had poor optical quality because the lenses had various defects. In 1729, the first lens that combined two different types of glass appeared. The word for this type of lens is "achromat," which means not color dependent. An achromatic lens does a good job of bringing all colors of light to the same focus.

In the 1980s, the first "apochromatic" lenses became available. An achromat is a two-lens system. Apochromats also may use two lenses, but they're more likely to have three or four. The main difference between the two types is the amount of "excess" color you'll see on bright objects. Excess color is not a color that's been added to what you're viewing. Rather, it usually appears as a purple fringe on one edge of the object.

Through an achromat, you'll see excess color on bright objects like Jupiter, Venus, and the Moon. Many observers can simply ignore this. Apochromats deliver an image essentially free of excess color. Because of the additional lens elements, however, apochromats cost more than comparable achromats.

Advantages of Refractors

High-quality refractors have a totally clear aperture (the size of the optic that collects the light). No central obstruction scatters light from bright to dark areas. That means image contrast is generally better in refractors. Observers of planets and double stars often cite refractors as the premier instruments for viewing those objects.

Fig. 14.3 Celestron's NexStar 102SLT package contains a 4-inch refractor on a computerized mount. (Courtesy of Celestron)

A second advantage of refractors is that they are low maintenance. Lenses never require recoating like mirrors do. Also, a lens system generally doesn't require adjustment, or what telescope-makers call collimation. The lens does not get out of alignment unless the scope encounters some major trauma. In other words, if you don't drop the telescope, you'll never have to align it.

Yet another advantage is that some of the newest refractors are among the smallest—and, therefore, most portable—telescopes made. Five decades ago, lens-making technology was not good enough to produce high-quality optics that focus in a short distance, making the instrument, in today's terminology, a short-focus refractor. To reduce the distortions observers would see, refractors sported lenses that focused images more than a meter away with correspondingly long tubes. Such instruments were hard to transport, and in use they didn't respond well to even moderate breezes.

The Disadvantage of Refractors

Because a refractor has a closed tube, it requires a certain amount of time to adjust to outside temperature when moved from a warm or cool house. Today's thin-walled aluminum tubes have reduced this period significantly, but you still have to take cool-down time into account.

Three things to know about refractors

- Refractors use a lens system to produce images.
- Refractors require the lowest maintenance of all telescopes.
- Many small refractors will mount on a sturdy camera tripod, making them the ultimate grab-and-go scopes.

Reflecting Telescopes

Scottish mathematician James Gregory invented the reflecting telescope and published a description in 1663. Although he's given credit for the invention, Gregory never actually produced the telescope.

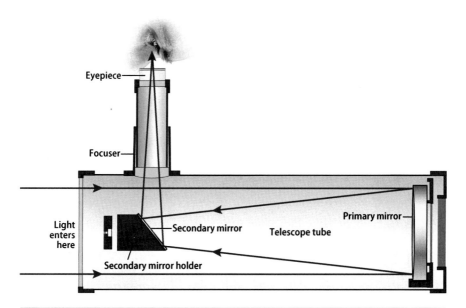

Fig. 14.4 A reflector uses a curved mirror to focus light and a small, flat mirror to reflect it to the eyepiece. (Courtesy of *Astronomy* magazine: Roen Kelly after Celestron)

English mathematician Sir Isaac Newton constructed the first working reflecting telescope in 1668. It had a mirror 1 inch across and a tube 6 inches long. A Newtonian reflector contains two mirrors—one called the "primary" at the bottom of the tube and a small, flat "secondary" near the top of the tube. Light enters the top, travels down the tube, hits the primary, and reflects to the secondary. That mirror then reflects it into the eyepiece.

Fig. 14.5 Celestron's SkyProdigy 130 is a reflector that contains a 5.1-inch primary mirror. (Courtesy of Celestron)

Early reflecting telescopes had mirrors made of solid metal polished to a high reflectivity. Unfortunately, metal tarnishes quickly so today's telescope mirrors are glass coated on the curved side with an ultra-thin layer of aluminum. Many are then coated to both enhance and protect their surfaces.

Advantages of Reflectors

Reflecting telescopes show no excess color. That means you won't see color fringes around even the brightest objects.

The biggest advantage of reflectors, however, is cost. When working with a mirror, manufacturers have to polish only one surface. An apochromatic lens has between four and eight surfaces, plus you're looking through the lenses so the glass has to be defect-free. Telescopes with apertures of more than 6 inches, with few exceptions, are all reflectors or compound telescopes.

Fig. 14.6 Celestron's AstroMaster 130EQ is a 5.1-inch reflector on an equatorial mount. (Courtesy of Celestron)

Disadvantages of Reflectors

The placement of the secondary mirror creates an obstruction that scatters a tiny amount of light from bright areas into darker ones. Unless you're looking at a planet or bright nebula under high magnification, you'll never notice this.

Newtonian reflectors suffer from "coma," a defect that causes stars at the edge of the field of view to look like a comet. Observers generally compensate for this by placing all targets at the center of the field of view.

Finally, because of how the mirror attaches to the tube, a reflector is sensitive to being bumped or jostled when transported. Advanced skygazers take no chances here. They usually collimate their telescopes before each observing session.

Three things to know about reflectors

- Reflectors use a system of mirrors to produce images.
- Reflectors offer the best "size per dollar" ratio.
- Reflectors are the largest amateur telescopes.

Catadioptric Telescopes

With regard to telescopes, catadioptric means "due to both reflection and refraction of light." These instruments also are known as compound telescopes and are hybrids that have a mix of refractor and reflector elements in their design.

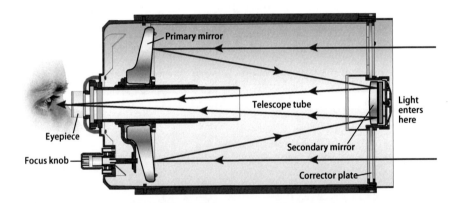

Fig. 14.7 A compound telescope combines a front lens with mirrors to focus light. This diagram shows a Schmidt-Cassegrain telescope. (Courtesy of *Astronomy* magazine: Roen Kelly after Celestron)

German astronomer Bernhard Schmidt made the first compound telescope in 1930. The Schmidt telescope had a spherical primary mirror at the back of the instrument and a glass corrector plate in the front that removed a defect called spherical aberration.

The Schmidt telescope was the precursor of today's most popular design, the Schmidt-Cassegrain telescope, or SCT. In this design, light enters the tube through a corrector plate. It then hits the primary mirror at the tube's base, which reflects the light to a secondary mirror mounted on the corrector. The secondary reflects light through a hole in the primary mirror to the eyepiece, which sits at the back of the scope.

Fig. 14.8 Celestron's NexStar 127SLT is a 5-inch compound telescope supplied with a go-to mount that runs on eight AA batteries (or an optional adapter). (Courtesy of Celestron)

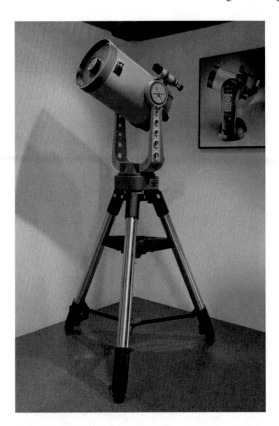

Fig. 14.9 Celestron's original C8 had an orange tube. (Courtesy of Celestron)

Advantages of Catadioptrics

The number one advantage of an SCT is its compact design. Such instruments are often only one-quarter as long as comparably sized reflectors and much shorter than refractors with half their aperture. This feature makes the SCT one of the ultimate grab-and-go telescopes.

Disadvantages of Catadioptrics

Like refractors, compound telescopes also have a closed tube. Adjusting to the outside temperature, therefore, takes longer than with an open-tube reflector of the same size. To compensate for this, Celestron installs filtered cooling vents behind the primary mirror of its top-end SCTs.

In 1970, Celestron began making a telescope that took amateur astronomers by storm: the Celestron 8, or the C8 as observers soon were calling it. The introduction of this scope resembled that of the first IBM personal computers a decade later— it started a revolution.

The orange-tubed Celestron 8 SCT had many advantages—8-inches of aperture, light weight, better portability than any 8-inch Newtonian reflector sold at the time, and an f/10 optical system, which provided good magnification. A range of ready-to-use accessories made astrophotography simple and popular. The complete system included a wedge (for polar alignment and tracking) and a sturdy, folding tripod.

Celestron based several of its current telescopes on this proven design. Among them are the CGEM, CGEM Edge HD, CPC, NexStar SE, and SGT lines.

Three things to know about compound telescopes

- Compound telescopes use both lenses and mirrors to produce images.
- Compound telescopes have the most compact design.
- Compound telescopes usually come as complete systems.

Mounts and Drives

We call our instruments telescopes, but the phrase "optical tube assembly on a mount" also works. In fact, it points out that half of any telescope is the mount. Some observers claim that the mount is the more important part. An unstable mount will render the finest telescope unable to deliver quality images. If the mount is undersized, wind—the bane of most large telescopes—will not be your only enemy. You will experience bouncing images even when you are focusing.

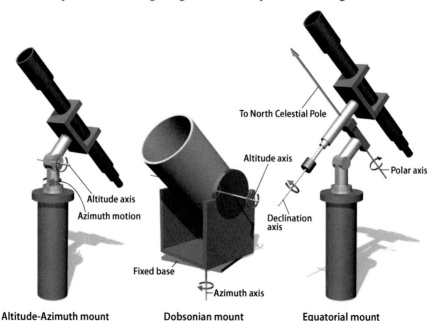

Fig. 14.10 The most popular amateur telescope mounts are shown in this illustration. (Courtesy of *Astronomy* magazine: Roen Kelly)

Alt-azimuth Mounts

An alt-azimuth mount is the simplest type of telescope mount. The name is a combination of altitude and azimuth. Altitude is the number of degrees a sky object lies above the horizon. Azimuth is the number of degrees from north to the object moving around the horizon. A telescope on this type of mount moves up and down (altitude), and left and right (azimuth).

Dobsonian Mounts

Telescope maker and amateur astronomer John Dobson invented a simple type of alt-azimuth mount that now bears his name. The Dobsonian mount is the least expensive telescope mount and manufacturers almost always combine it with a Newtonian reflecting telescope. Because the tube sits loosely in the mount, an observer can carry the two parts of a small Dob (as observers call them) quite easily. But they also can be made large. Every amateur telescope that has a mirror more than 16 inches across sits in a Dobsonian mount.

Fig. 14.11 Celestron's AstroMaster Tripod is a simple alt-azimuth assembly on which you can mount binoculars or a small telescope. (Courtesy of Celestron)

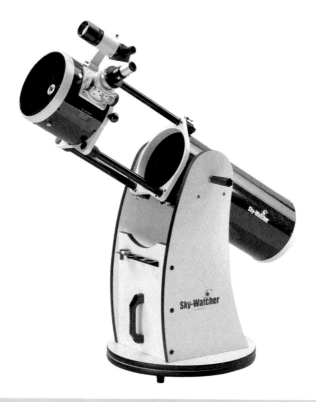

Fig. 14.12 Celestron's Sky-Watcher Dobsonian line combines a Newtonian reflector with an easy-to-use Dobsonian mount. (Courtesy of Celestron)

Go-to Mounts

A development in mounts that occurred in the 1980s is the *driven* alt-azimuth mount, also called a go-to mount. To create this, manufacturers attach motors to both the altitude and azimuth axes. The motors also interface to a computer. Once you run through a simple setup procedure by centering one or two stars for alignment, you're ready to observe. Simply enter the object's name or number, or select it from a list contained in the telescope's hand controller, and the go-to drive first finds your target and then continues to track it.

Mounts employing this system are highly accurate. Once the drive locates an object, you'll be able to follow it without moving the telescope assuming that your initial alignment was a good one. As noted, most go-to telescopes manufactured today have a large, built-in database of objects from which you can select.

Fig. 14.13 Celestron's NexStar 6SE package combines a 6-inch Schmidt-Cassegrain telescope, a tripod, and a computerized go-to mount. (Courtesy of Celestron)

Equatorial Mounts

If Earth did not move, a non-motorized alt-azimuth mount would be all that any of us would ever need. But our planet does spin, and astronomers must take it into account. The second type of mount is the equatorial mount. German optician Joseph von Fraunhofer invented it in the early 19th century to track the apparent motion of the stars. He succeeded by aligning one of the mount's axes parallel to Earth's axis and by moving the mount with a weight-driven clock drive at the same rate as our planet's spin. Today equatorial mounts use a motor to move them. With the addition of a computer and a database, an equatorial mount also becomes a go-to mount.

Three things to know about mounts

- A mount defines how a telescope moves.
- A mount is as important as the optics.
- A go-to mount will enhance your observing.

What About Magnifying Power?

Avoid any telescope box that touts magnification. Claims of 500× for telescopes are meaningless, and here's why: All you need to do to change the magnification of a telescope is change the eyepiece. So, if a low-quality, high-power eyepiece comes with the telescope, the instrument can achieve high albeit meaningless magnifications.

Normally, amateur astronomers calculate a telescope's *maximum* useful magnification by multiplying the size of the lens or mirror in inches by 50. This means a 4-inch telescope can be used with eyepieces that provide up to about 200×. An 8-inch telescope's highest useful magnification is 400×, and so on. Take it from long-time observers—you'll use low magnifications in your telescope far more often than high ones.

Finding Magnification

To calculate the magnification, or power, of any eyepiece, simply divide the telescope's focal length (listed in the instruction manual, on the tube, or on the front lens of some refractors and catadioptric scopes) by the eyepiece's focal length (the number printed on the eyepiece). For example, let's say that a certain telescope has a focal length of 1,000 millimeters. If you choose a 25 millimeters eyepiece, the magnification will be 1,000 divided by 25, or 40×. If you replace the 25 millimeters eyepiece by a 10 millimeters eyepiece, the magnification will be 1,000 divided by 10, or 100×.

Try Before You Buy

If you are serious about purchasing a quality telescope, try to observe through any model you're considering. The easiest way to do this is to call or visit a local astronomy club. Find out when they're hosting a public observing session and attend it. Observe through as many telescopes as you can, and ask as many questions as you can think of. Ask about setup time, maintenance, accessories, and cost. Also, especially if you visit a nighttime observing session, ask about the potential daytime use of any telescope you're interested in. Has the owner observed the Sun through it? What kind of filter did they use? If you can't find an astronomy club nearby, call or visit a local planetarium. The staff there will be aware of any astronomy clubs in your area.

Fig. 14.14 An astronomy shop might let you look through the telescope you're thinking of buying. (Courtesy of Celestron)

Seek Out a Reputable Dealer

Telescope dealers that have been in business a while have sold scopes to all types of buyers. Long-time dealers will ask you questions to determine your level of expertise and your interests. A good dealer will not try to sell you more telescope than you need. For this reason, I cannot suggest that you purchase a telescope from a department store or catalog showroom.

Top 10 things you should know before buying your first telescope

10. Three main types of telescopes exist.
 9. The mount is as important as the optical tube.
 8. A go-to drive will enhance your experience.
 7. You can "test-drive" telescopes.
 6. Make sure your telescope has what you need.
 5. Do some research.
 4. Set up your telescope in the daytime.
 3. You get what you pay for.
 2. Bigger means better.
 1. Pick a telescope you'll use.

Telescope Accessories

Once you have a telescope, you can turn your scope into a complete system by adding some well-thought-out extras, listed below.

Finder Scopes

The best telescope in the world is useless if you can't find anything through it. Even with a go-to drive, you'll need a quality finder scope. Most are straight-through finders, which flip the image but let you sight down the main telescope tube, a position that's intuitive for most people. Right-angle finders keep the field of view correct, but looking at the telescope trying to find objects is difficult for some observers.

Fig. 14.15 Celestron's Finderscope kit features a finder scope with a 2-inch (50 millimeters) front lens and a magnification of 9×. (Courtesy of Celestron)

Fig. 14.16 This small finder scope doesn't magnify. Rather, it projects a red dot onto a transparent screen. (Courtesy of Celestron)

Your finder should have a front lens at least 2 inches (50 millimeters) in diameter. That size will let enough light in so you won't get frustrated trying to find faint objects. The finder's magnification should be between 7× and 9×.

Once you install your finder scope, align it with your telescope. Do this when it's still light. It's easier then because the objects you'll use to align your finder won't be moving like the stars do.

IMPORTANT ECLIPSE-RELATED NOTE: If you will be using a telescope to observe the eclipse, either remove your finder scope, cover it, or purchase an approved solar filter for it, and secure it so it won't fall off (or be easily removed). That way, neither you nor anyone else will mistakenly look through the finder at the Sun's full glory. To say that would be bad is an understatement.

Fig. 14.17 A star diagonal bends light 90°. This accessory makes observing a lot more comfortable. (Courtesy of Celestron)

Finder scope setup

Align your finder scope before each observing session while it's still light. Here's how:

- If your telescope has a motorized drive, turn it off.
- Insert a low-power eyepiece (the one with the biggest number).
- Loosen your drive's motion-control locks.
- Move your telescope until you center a distant earthly object (the light on a transmission tower, a building, etc.). Focus your scope on the object.
- Lock your telescope's motion controls.
- Loosen the screw locks on your finder scope's mounting bracket and then (without moving the main scope) position the finder scope so the object you centered also is centered in your finder. Note that, depending on the type of finder scope you have, one of the images may be upside-down and the other correctly oriented. Focus on a specific spot, not the whole image.
- Lock your finder scope into position.
- For higher precision, replace the low-power eyepiece in your telescope with a high-magnification one, then realign.

Star Diagonals

Refractors usually need a star diagonal because of their design. A star diagonal bends the light from your target 90° into the eyepiece. Without a star diagonal, you'll find yourself in some awkward positions when you're observing objects high in the night sky or when you're looking at the Sun on eclipse day. The star diagonal fits into the telescope's focuser, and the eyepiece fits into the star diagonal.

Fig. 14.18 Celestron's PowerTank 17 is a 12-volt power supply that will run your telescope and other accessories when you're out observing at a site without electricity. (Courtesy of Celestron)

Power Supplies

If you observe from a location with alternating-current power, consider yourself lucky. The rest of us need some form of portable power. With the right adapter, you can use your car's battery.

Another option is a dedicated power supply. Celestron's PowerTank 17, for example, has plenty of power for several all-night sessions. It also includes an AM/FM radio, a siren, a removable red-filtered flashlight, and a white spotlight.

Camera Adapters

Someday, your desires may turn to astroimaging. But what if you don't have a digital single-lens reflex camera with the correct adapter? Celestron and other manufacturers make a universal mounting platform you can use with a point-and-shoot digital camera to photograph what you see through the eyepiece.

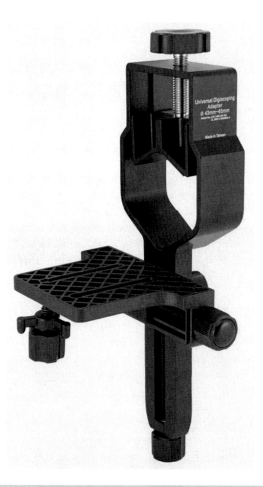

Fig. 14.19 Celestron's Digital Camera Adapter allows you to photograph through your scope's eyepiece. (Courtesy of Celestron)

Fig. 14.20 A solar filter like this one, which is made of glass, will let you view the Sun anytime. Solar filters always go on the front end of a telescope. (Photo courtesy of the author)

Moon Filter

This specialty filter sometimes goes by the name neutral density filter. It reduces the amount of light by absorbing it but doesn't filter any of the colors. Neutral density filters let as much as 80 percent and as little as 1 percent of the light through. In general, lighter neutral density filters are used for the planets and darker ones for the Moon, which reflects much more of the Sun's light. *No* neutral density filters are safe to use to view the Sun. You should use *only* filters that cover the front end of your telescope.

Using a Telescope for the Eclipse

Assuming that you already own a telescope or that you've just purchased your first one, here are some suggestions for observing that will prepare you for August 21, 2017.

Obtain a Solar Filter

In addition to using your telescope to view the eclipse, you can double your potential observing time if you purchase a quality solar filter and observe the Sun. As mentioned above, a filter that fits over the front of your telescope is the only kind to use. *Never use a filter that screws into an eyepiece to view the Sun.* Be absolutely certain wind or accidental bumps cannot dislodge your filter. If you are in doubt, securely tape the side of the filter to the telescope tube.

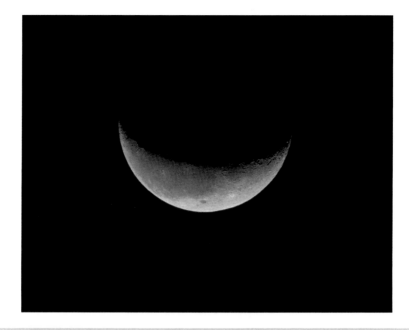

Fig. 14.21 To practice for the August 21, 2017, total solar eclipse, observe the Moon through the same setup you'll use to view the Sun. (Courtesy of NASA)

Start your solar observing by counting sunspots. Apart from being fun, sunspot counts let you know how active the Sun is. And if you watch or especially sketch from day to day, you'll be able to see that the Sun rotates. People have been recording sunspot numbers on a daily basis since the middle of the 18th century.

For a discussion about the different kinds of solar filters available, see Chapter 16.

Observe the Moon

The Moon offers something for everyone. It has a face that's always changing. Following it telescopically through a lunar month can be fascinating. But when the Moon is brightest, at Full Moon, is the worst time to view it. From Earth, the Sun is behind us, so there are few shadows and we see little detail.

The best evening viewing times are from when you can just see the thin crescent after New Moon until 2 days after First Quarter. In the morning, view from about 2 days before Last Quarter to just before New Moon. Shadows are longer at these times, and features stand out in sharp relief.

Fig. 14.22 Celestron's 18-millimeter X-Cel LX eyepiece offers medium magnification in most telescopes. (Courtesy of Celestron)

Concentrate looking along the Moon's shadow line, called the terminator. It divides the light and dark portions. Along the terminator, you'll see mountaintops protruding high enough to catch sunlight while dark lower terrain surrounds them. On large crater floors, you can follow "wall shadows" cast by sides of craters hundreds of feet high. All these features seem to change in real time, and the differences you can see in just one night are striking.

That may be all well and good, but how does a discussion, albeit a short one, about the Moon help you prepare for the eclipse? Merely observing the Moon, which has more visible detail on its surface than any other celestial object, will make you a better astronomer. What's more, and this is especially true if you don't have a solar filter yet, recall that the Sun and the Moon appear the same size in the sky. Viewing the Moon through your telescope will familiarize you with how a Sun-sized object looks through your telescope through different eyepieces.

All About Eyepieces

Eyepieces are like stereo equipment. In both cases, people recognize and admire high-quality workmanship. You want a sound system that faithfully reproduces music as close to the original as possible. And yet, while listening to a familiar piece of music, each of us perceives something a little bit different about it. You may hear some nuance meaningful to you that I didn't catch. The end result is that we don't all end up with the same stereo equipment, or eyepieces.

Fig. 14.23 Celestron's Eyepiece and Filter Kit contains five eyepieces, a Barlow lens, six color filters, and a Moon filter. (Courtesy of Celestron)

Sometimes this is due to cost. The best eyepieces contain multiple highly polished and coated lenses made from exotic glass, so they are not cheap. Coatings, by the way, are ultra-thin layers manufacturers apply to lenses to reduce the amount of light reflected by the eyepiece's glass and thus increase the amount that passes through.

Some hobbyists find it tough to justify spending as much on a few eyepieces as on their telescope. Most amateur astronomers, however, look at the investment over the long term. If you upgrade your telescope, you don't need to change your eyepieces.

When choosing which eyepiece to buy, consider its weight. Some tip the scale at more than 2 pounds—as much as some binoculars. If you purchase a small or medium-sized telescope, you'll want to choose lighter eyepieces.

Fig. 14.24 Celestron's Omni Barlow lens doubles the magnification of any eyepiece. (Courtesy of Celestron)

Another thing to keep in mind is the eyepiece's field of view. You'll see two numbers used, the apparent field of view and the true field of view. The apparent field of view of an eyepiece is the angle of light able to enter the eyepiece. Eyepiece apparent fields range from 25° to 84°. An eyepiece's true field is the amount of sky you actually see when you look through the eyepiece. This number can change from one telescope to the next.

Barlow Lenses

A Barlow lens is an optical accessory that increases an eyepiece's magnification. It's a tube with a few lenses that goes between the telescope's focuser—or the star diagonal if you're using one—and the eyepiece. Some Barlows magnify two times (2×), some are 3×, and so on. So, as an example, let's say your 18 millimeters eyepiece gives a magnification (you'll also hear this called "power") of 100× through your telescope. If you insert a 2× Barlow, the magnification will be 200×.

Fifty or so years ago, when Barlow lenses first appeared, they were simple units using single lenses. They worked, but they worsened the view. Today's Barlows contain high-quality coated lenses with great light transmission.

A Barlow lens can effectively double the number of eyepieces in your set if you select your eyepieces with this in mind. Here's an example: Let's say you have 40, 32, 12, and 9 millimeters eyepieces that, in your telescope, magnify 25×, 31×, 83×, and 111×, respectively. Adding a 2× Barlow lens will give you four additional magnifications: 50×, 62×, 166×, and 222×.

Suggestions for the Eclipse

Your first thought after purchasing a new telescope should be to get an approved solar filter for it. Nothing is more important than this.

Once you've accomplished that, the overriding concern you should have is the selection of eyepieces. Be sure when you select a telescope and an eyepiece to view the eclipsed Sun through that the combination produces a *true field of view* (TFOV) wide enough for your desires. The Sun and Moon are each approximately 0.5° wide. Obviously then, you won't be choosing a telescope and eyepiece combo whose TFOV is less than that.

When I've used a telescope to view solar eclipses, I carefully select just two eyepieces to combine with it. The first one gives a TFOV of 1°. That makes the Sun's disk half as wide as the whole field of view my eye sees when I look through the eyepiece. The second eyepiece I select provides a TFOV of 2°, which makes the Sun one-quarter the diameter of the field.

Each eyepiece has a specific purpose. I select the first one (1° TFOV) for the partial phases of the eclipse. This makes the Sun large enough for me to see any sunspots on its surface. As totality approaches, however, I switch to the 2° TFOV eyepiece because I want any views through the telescope to show a good part of the Sun's corona, which appears as a ring around the eclipsed disk. Simple math shows that the ring around the Sun you'll see through the higher-magnification eyepiece will be ¼° wide. The lower-power eyepiece, however, will produce a ring around the Sun that's ¾° wide. If I want a wider view yet, I switch to binoculars.

The TFOV of a telescope/eyepiece combination is calculated by dividing the eyepiece field stop diameter by the telescope's focal length and then multiply that result by 57.3. The field stop is the ring inside the eyepiece barrel that limits the size of the field of view. The eyepiece projects a virtual image of it, which appears as a circle when you look through the eyepiece. Some manufacturers provide the eyepiece's field stop in the specification sheet that comes with the eyepiece. Most do not.

But don't despair if that information didn't come with the eyepiece because there's another way to calculate TFOV that's only slightly less accurate. Divide the eyepiece's apparent field of view by its magnification when it's in your chosen telescope. To get this result, you first have to calculate the magnification, which I discussed above. Then you divide that number into the eyepiece's apparent field of view, a number provided by almost all manufacturers. Let's look at an example.

For the upcoming eclipse, you want to see if your favorite scope will work. It's small, and its mount and tripod aren't huge either. That makes it easy to transport in your compact car. The scope has a focal length of 500 millimeters, and you want to see what the TFOV will be with your 25 millimeters eyepiece, which has an apparent field of view of 50°:

$$\text{Magnification} = \text{focal length of telescope} / \text{focal length of eyepiece}$$

$$\text{Magnification} = 500 / 25 = 20 \times$$

$$\text{TFOV} = \text{apparent field of view} / \text{magnification}$$

$$\text{TFOV} = 50° / 20 = 2.5°$$

For this telescope/eyepiece combo, your TFOV would be 2.5°. That would make the diameter of the Sun's disk 20 percent as wide as the view through the eyepiece, and it would leave a 1° ring around it. Not too shabby!

Then you start thinking about a second eyepiece. How about that nice 10 millimeters? It also has an apparent field of view of 50°. In this case,

$$\text{Magnification} = 500 / 10 = 50 \times$$

$$\text{TFOV} = 50° / 50 = 1°$$

This would be an excellent duo of eyepieces to mate to the telescope described above.

Allow me to share one more piece of advice. Keep things simple. Don't bring more than two eyepieces for eclipse viewing. This will make the transition from your high-power to low-power views foolproof. You won't grab the wrong eyepiece or be looking down to be certain you get the right one. Remember: Time is of the essence!

Chapter 15

Pick the Right Camera for the Eclipse

Hopefully by now you know where I stand on first-time eclipse viewers trying to photograph the spectacle. If you don't know, I am dead-set against it. For the vast majority of viewers it will dramatically cut into the time that should be spent on viewing the event, and trying to capture the eclipse can ruin the experience, all for that one-in-a-million shot that is unlikely to make it into *Astronomy* magazine.

On a much more positive note, I am that magazine's photo editor, so I really can help you figure out what features to look for if you'll be purchasing a camera for your effort. But first, let me save many of you some time. For best results, you must shoot through a digital single-lens reflex camera (DSLR). If you plan to photograph the eclipse with your phone or with a point-and-shoot camera, you can stop reading this chapter now. Those devices will allow you to capture a memory, but little else. The graphic below shows why.

Which Sensor?

Having a larger sensor (often called the camera's chip) is one of the main reasons you'll get much better results when you shoot the eclipse or anything else through a DSLR. Even with a DSLR, however, you have two choices when it comes to the sensor inside. One is the Advanced Photo System type-C (APS-C) sensor, which

© Springer International Publishing Switzerland 2016
M.E. Bakich, *Your Guide to the 2017 Total Solar Eclipse*, The Patrick Moore
Practical Astronomy Series, DOI 10.1007/978-3-319-27632-8_15

Fig. 15.1 This graphic shows the exact ratios of chip sizes in various cameras. The *blue rect-angle* represents a full-frame DSLR chip. The *yellow* shows the APS-C sensor. The *red* is the average size of the sensor in a point-and-shoot camera, and the *black rectangle* shows the size of a cellphone chip in comparison. (Courtesy of Holley Y. Bakich)

has an image ratio of 3:2. This mimics the Advanced Photo System format developed by several film manufacturers in 1996. The size of the "C" film was 25.1 by 16.7 millimeters, which is a 3:2 ratio.

The second choice is a full-frame sensor, which measures 36 by 24 millimeters, the same size as 35 millimeters film. The first successful commercial full-frame DSLR was the Canon 1D, introduced in 2002. These days, professional photographers demand nothing less than full-frame sensors because they collect more light than their smaller counterparts, and that produces a better signal-to-noise ratio. This translates to a better overall image, especially if you shoot at high ISO numbers.

The larger sensors also capture more of an image from whichever lens you choose while smaller ones crop in on the image, producing a view with a narrower angle. As an example, let's say you take a picture through a full-frame DSLR with a 20 millimeters lens. If you took the same shot using the same lens through a camera with an APS-C sensor, the view would appear like you used a 31 millimeters lens on the full-frame DSLR. One benefit, then, of the larger sensor is that it makes wide-angle photography easier.

The other main difference between DSLRs with the two types of sensors is that full-frame models tend to be more expensive—much more expensive. When we consider the upcoming solar eclipse, is the extra expenditure worth it? No, it's not, in my opinion. All things being equal, you will get images from APS-C sensors that will blow your mind. Why spend $2,000 or more for a camera (just the body, now, we're not adding in lenses yet) when a $500 model will do?

More Megapixels?

Manufacturers measure camera resolution in megapixels, which equals 1 million pixels. Each pixel is a light-collecting element on the camera's sensor, so it's also the smallest unit of a digital photo. What then makes a camera a 12-megapixel model? If its chip contains rows 4,000 pixels wide and columns 3,000 pixels high, then $4,000 * 3,000 = 12,000,000$ pixels, or 12 megapixels.

Because of all the advertising on television and in magazines, you probably think that a sensor with more megapixels will deliver better resolution, and, therefore, a superior image. That would be true only if all pixels were the same. They're not.

Ads for cellphones often tout huge numbers of megapixels. Forty million pixels on a sensor? Wow! But, wait a minute. As stated above, cellphone chips are roughly 2 percent the size of a full-frame sensor. To pack those millions of pixels onto such a small sensor, camera phones, and even point-and-shoot cameras, use much (much!) smaller pixels than DSLRs. Unfortunately, small pixels pretty much suck. The larger pixels used on full-frame or APS-C sensors capture more light, produce better color, and have less noise.

The Camera's Processor

If you're a beginning photographer, or even one with some experience who will be upgrading your camera, having a high-quality image processor will help you a lot. If you're more experienced, and especially if you shoot images in RAW format, the internal processor won't matter much to you. RAW images are just the data the sensor sees, completely unprocessed. This setting is for those of you who want to process the images yourself using software like Photoshop.

There's one other reason people shoot JPEGs rather than RAW images: JPEG is a compressed format, meaning that one takes up a lot less space on your camera's memory chip than a RAW image. If, however, you're like most people and shoot JPEGs, the processor matters. Processors that produce JPEGs handle operations that can fix lighting issues and adjust various other settings. It also lets the camera capture images in quick succession. While this doesn't affect image quality directly, being able to capture images quickly can mean the difference between a good shot and a great shot.

Lens Differences

The camera's lens is its eye to the universe. If it doesn't perform well, you will not be happy with your images. Cellphones and point-and-shoot cameras won't produce high-quality images because their lenses are tiny and fixed to the camera. Manufacturers measure lenses by their aperture, which is how wide the lens can open. A wider aperture means more light. More light means you can take photos more easily where there isn't a lot of it.

"But Michael," you may ask, "aren't we dealing with the Sun here? There will be plenty of light, right?" Not during totality. The illumination during the total phase of the eclipse will be only one-millionth what it was when the disk was visible. So, indeed, you do need to factor in a lens' light-grasp.

I provide specific lens considerations in Chapter 17.

Eclipse Considerations

Which DSLR should you buy with the eclipse in mind? Before I answer that, let me tell you where you must look: eBay. I dialed up the giant online auction house and entered a few criteria. Frankly, I was amazed at the quality of camera body you can get for under $500 (my self-imposed upper limit for this chapter), and some cost quite a bit less than that mark. A bit fewer than 50 percent were new, and the others were used but seemingly in immaculate condition. And I was just surfing the "Buy It Now" options. You might be able to get a screaming deal if you select "Auction."

I selected "Canon" as my manufacturer, but "Nikon" will work equally well. I'm a lot more familiar with Canon cameras, plus more than 90 percent of the images I receive at *Astronomy* magazine originate in that brand. But I'm also familiar with Nikon quality. These manufacturers are the two giants, so I suggest you look no further and go with one of them.

And like I mentioned in the "Which sensor?" section above, get a camera with an APS-C sensor, although that's pretty much the default here. You won't find a new or used full-frame sensor DSLR for under $500.

As far as the imaging chip, set 12 megapixels as your minimum. That will produce images with dimensions of 4,000×3,000 pixels, which will allow you to zoom in for your final composition, if that's what you want. Heck, I saw Canon EOS Rebel T5 bodies for less than $300, and they sport an 18-megapixel sensor.

With regard to lenses, please read through Chapter 17 carefully. You'll find all the calculations you'll need to help you determine what lens you need for a specific task.

Finally, if you will be shooting through a telescope, remember that you'll need a T-adapter and a T mount. See item #13 in Chapter 17 for more about this.

Chapter 16

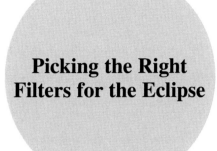

Picking the Right Filters for the Eclipse

Although viewing our local star can be dangerous, a number of safe methods exist. These include handheld and telescope-mounted filters. And while I called the first category "handheld," it includes both welder's filters and solar viewing glasses.

Welder's Filters

For most eclipse-chasers of a certain age, the original handheld filter was either a homemade one that they built out of cardboard plus a piece of solar Mylar or a #14 welder's glass. Currently, welder supply stores generally do not stock #14 filters. That number is about the darkest ever made, and the only "job" use I could find for it was to protect someone's eyes when they were doing carbon-arc welding. You still can order them, however, either at a shop or online, at a cost of only a few dollars.

Most people don't like them because they impart a greenish hue to the Sun. I guess that added color doesn't bother me because I have been using the same 4¼″ by 2″ piece of #14 welder's glass since my first total solar eclipse in 1970. My wife, Holley, agrees. She's used #14 for the six totals she's seen and has no problem with the view.

© Springer International Publishing Switzerland 2016   163
M.E. Bakich, *Your Guide to the 2017 Total Solar Eclipse*, The Patrick Moore Practical Astronomy Series, DOI 10.1007/978-3-319-27632-8_16

Fig. 16.1 Number 14 welder's glass is a safe solar filter. You can view the Sun through it any time. Note that tests have shown #12 welder's glass also to be safe, however, many people find the Sun's image through a #12 welder's filter uncomfortably bright. (Photo courtesy of the author)

That said, for most people the best way to go is to select a model from the new crop of eclipse glasses. These devices have several advantages over welder's glass. The glasses made of cardboard and solar Mylar are much lighter. You don't have to hold glasses to view the Sun—you wear them. And although you can slip your welder's glass into goggles designed to hold them, that combination is heavier and bulkier. Finally, if you happen to drop each of these on a hard surface, more than likely the glasses will survive intact.

One other note about welder's glass before I move on. Some of the new models are not glass. Rather, some manufacturers have started making welding filters out of a polycarbonate material. A year ago, nobody in the know would have recommended using one of them. Everyone thought that while they blocked the visible light, such filters transmitted an unsafe level of near-infrared radiation. But as of June 2015, a new standard (ISO 12312-2:2015) has emerged as the international guide to safe solar eclipse viewing.

Commenting on it, one of its architects, B. Ralph Chou, Professor Emeritus, School of Optometry and Vision Science, University of Waterloo, said, "The essential part of the standard is that solar filters are to have luminous transmittance equivalent to between Shade Number 12 and 16, and there are additional constraints on UV and IR transmittance, but these are not the same as the transmittance requirements for welding filters. In our rationale for the transmittance levels, David

Fig. 16.2 Rainbow Symphony's solar viewing glasses may resemble standard sunglasses, but the *only* thing you can see through them is the Sun. (Courtesy of *Astronomy* magazine)

Sliney and I wrote that the UV and IR levels in sunlight are not a significant factor for solar retinopathy—the injury is primarily due to short-wavelength visible light."

This means it's OK to use a polycarbonate welder's filter if you can find one with a dark enough shade (#14), though some differences do exist between them and glass versions. Glass welding lenses are available in more styles and shades, so your selection is better if you go with glass. Glass lenses can break. Polycarbonate lenses will never break. Polycarbonate lenses scratch much easier than glass ones, so glass lenses will look much better over time, provided you don't drop them. Finally, glass welding lenses have reasonably good optical quality (as good as a ¼″-thick piece of glass can have); polycarbonate's optical quality is significantly lower.

Solar Viewing Glasses

The easiest and cheapest way to view the Sun is through solar viewing glasses. These devices incorporate a cardboard frame to hold an approved filter, usually solar Mylar or a black polymer material. The advantage of the polymer is that it holds up to use (and unintentional abuse) better than Mylar. One of the two large suppliers, American Paper Optics, claims to have sold two billion eclipse glasses in the past quarter century. In even small quantities (25), the price per unit is less than $1.

The other main retailer, Rainbow Symphony, recently introduced a line of designer eclipse glasses. Rather than a thin cardboard frame, the new models had frames of molded plastic—in three colors. Some were "normal" glasses, and others used the wrap-around style. All are safe for solar viewing. In fact, you can't see anything *but* the Sun through these glasses.

The one potential issue with the designer glasses is that someone less informed than you may think you're casually watching the Sun through sunglasses, a real no-no. If anyone's around when you're stylin' with the Rainbow Symphony look, make sure to take the opportunity to tell him or her about the eclipse and about safe solar viewing. If you're really feeling social, you can let them try on the glasses and have a look at old Sol himself.

Telescope Filters

The most important thing about using a filter with a telescope is that the filter *always goes on the front end of the scope*. Never use any filter that fits over or screws into an eyepiece. They have been known to crack, and if you're looking through one when that happens, the result could be devastating.

The oldest option is a metalized glass filter that uses flat polished glass coated with aluminum, nickel, or chromium to drop the Sun's brightness to safe and comfortable levels. Most glass filters impart an orange color to the image.

Another type of Sun filter for your telescope is made of solar Mylar mounted in a metal or plastic cell. Usually these cost a little less than glass filters. Mylar filters make the Sun appear slightly blue, which some observers don't like.

A third type of filter uses the "solar safety film" developed by Baader Planetarium in Germany. The filter material is a high-strength polymer metallized on both sides. Baader astrofilm provides a white image of the Sun.

Don't be concerned if you see wrinkles in the Mylar or Baader filters. Because your telescope isn't focusing on them, the image quality isn't affected. A "stretched flat" film filter is more likely to tear from even small impacts.

Fig. 16.3 The author constructed this filter by combining a hardwood frame and Baader AstroSolar Safety Film. The unit fits snugly over the front end of a telescope, and the black knob (attached to a threaded brass rod) secures it. (Photo courtesy of the author)

Finally, Rainbow Symphony has recently introduced a line of black polymer filters designed to fit binoculars, telescopes, camera lenses, and even finder scopes.

All of these filters work well, but they come in two different types: full-aperture and off-axis. Observers prefer full-aperture models, especially for refractors. That's because aperture is one of two factors that determines how much detail you'll see in a celestial object. (The other factor is seeing, a measure of the steadiness of the atmosphere.)

If your telescope's aperture is larger than 8 inches—and pretty much 100 percent of these are reflectors or catadioptrics—you'll want to use an off-axis filter for two reasons. The first is cost. The second reason is that an off-axis filter will use the part of your aperture not obstructed by your scope's secondary mirror holder.

Cool Products for a Hot Sun

As a staff member at *Astronomy* magazine, I have access to all kinds of new products. In that process I reviewed five terrific products that have given who-knows-how-many people high-quality views of the Sun. Two are filters you add to existing telescopes, two are stand-alone solar scopes, and one is … different. In addition to writing about these for the magazine or editing pieces on them, I currently own, or have owned, every piece of equipment listed below. They offer tailored options beyond the traditional welder's glass or solar glasses.

Solarscope

While this section deals with solar filters for telescopes, there is another product, similar to a telescope, which in this case projects an image. That item is a Solarscope. Set up this simple solar viewer during the partial phases of the eclipse, and you're sure to attract a crowd. It's an ideal product for Sun-viewing that combines safety and low cost.

Fig. 16.4 The Solarscope. (Courtesy of *Astronomy* magazine)

The Solarscope was invented by astronomer Jean Gay from Cote d'Azur Observatory in Nice, France, as an easy way for groups to observe the Sun. The product is light-years ahead of a cardboard box that uses pinhole projection, but it stops just short of being a telescope. It's also safer than projecting through a telescope, since inexperienced observers may look through the scope, and it's less expensive than using many other filters.

The viewer comes in two sizes for either one person or a group and is made of sturdy cardboard. The user must construct it themselves, but a booklet shows how the parts fit. As with a pinhole-projection system, you need a shaded viewing area. With the Solarscope's cleverly designed base and hood assembly, the Sun's image projects onto a dark area, which makes it easy for group to view.

Solarscope's optics are pretty clever. During assembly, you insert a mirror into an aluminum holding device. This snaps into the base of the Solarscope. Because of the mirror's shape, the Sun's image is offset inside the box. The other optic is a simple "telescope" made of a plastic tube with a front lens, but without an eyepiece. The telescope attaches to the hood. To focus the Sun's image, simply screw the telescope in or out.

Once you fit everything together, you're ready to observe. It takes about 10 minutes to assemble the Solarscope, then using it a piece of cake. Just set the assembled instrument on a table or a stand. In a pinch, you can set it on the ground, but raising it up makes the Sun easier for people to see. To point it, move the hood with the telescope up or down while turning the base side to side until the telescope points at the Sun.

Fig. 16.5 A group of people using the Solarscope can view the Sun's projected image. (Courtesy of *Astronomy* magazine)

Once you've found the Sun, the box's interior lights up. Keep moving the unit until the beam of sunlight hits the mirror. Rotate the mirror slightly to throw the Sun's image to either side of the telescope. The Solarscope shows even medium-size sunspots.

The lens aperture measures 1.6 inches (40 millimeters). Solarscope's Individual model projects an image of the Sun approximately 3.2 inches (80 millimeters) in diameter. The Education model projects on approximately 4 inches (100 millimeters) across. Finally the screen sizes for the Individual and Education models are 9.4 inches (240 millimeters) square and 14 inches (356 millimeters) square, respectively.

Watching the Sun move across the inside is a nice way to demonstrate Earth's daily motion. Eventually, however, Earth will rotate enough that you have to realign the Solarscope on the Sun. Simple adjustments toward the west and either up or down are all that's required. One benefit is that if teachers or other group leaders want to use multiple Solarscopes, users can operate them without fear of eye damage.

The Hydrogen-alpha wavelength

The most popular narrowband solar-observation wavelength is Hydrogen-alpha (Hα). The Hα emission line occurs when a hydrogen atom's electron makes a transition from a higher energy level (the third) to a lower one (the second). The atom emits this energy at 6,562.8 angstroms (Å). This allows astronomers to view solar features like prominences and the chromosphere, which glow at this wavelength but are not normally visible. Visually, our eyes detect light in a range from 4,000 Å (violet) to 7,000 Å (deep red). But to view the Sun in Hα light takes a special filter.

Solarscope maintains a web site at www.solarscope.com that provides helpful ideas, information, and projects. For a small investment, here's a great way to help others view the Sun. This clever instrument gets two thumbs up from me. It may be made of cardboard and plastic, but it should have a place right next to your telescope.

Daystar Quark

If you plan on using binoculars or a telescope during the great solar eclipse of August 21, 2017, you'll be one of lots of people who will purchase an approved filter. Most will buy visual solar filters, also known as "white-light" filters. Such an accessory reduces the Sun's visible light to a safe and comfortable level and lets you see sunspots.

But if you want to see more than just sunspots, you'll need to move into the realm of Hydrogen-alpha (Hα) observing. Daystar Filters, a company that has been making such products for decades, has introduced a new player—the Quark—into that arena. It combines an all-in-one unit and the ultimate in portability with a price point below all but the smallest Hα filters.

Fig. 16.6 Daystar's Quark is a Hydrogen-alpha filter that fits into your telescope's focuser. The unit weighs only 13.9 ounces (394 grams). (Courtesy of *Astronomy* magazine)

At first glance, you might mistake the Quark for a sleek 2″ eyepiece. Daystar gave this product a cool look by anodizing the aluminum body black and red. The company also maximized its use and sales by making it fit both 2″ and 1¼″ focusers. The filter outputs to a 1¼″ eyepiece or camera adapter using a brass compression ring to avoid marring the barrel, and the company does sell a 2″ eyepiece holder for $45. To use it, you just unscrew the 1¼″ one and screw in the larger piece.

Daystar designed the Quark to work on refractors with focal ratios from f/4 to f/8, which immediately puts it in play with "grab-and-go" scopes. The filter has an internal heater that requires power to operate, and you can get that power in two ways: either from a USB 3.0 port (5 volts, 1.5 amps) or via the 90–240 VAC wall adapter (that even comes with international plug adapters).

If you plan to do a lot of observing away from home, it would also be wise to purchase the 30-amp battery pack (QBP30) that Daystar offers for $89. This unit has both 2-amp and 1-amp USB outputs (the Quark filter uses the 2-amp one) and a solar panel that recharges it on site. Because the Quark requires 1.5 amps, a fully charged battery pack will run it all day, even without the solar-panel recharging.

Since no single filter is perfect for all applications, Daystar produces two versions of the Quark: Chromosphere and Prominence. As the names indicate, each has its specialty, although there's enough overlap that you will see prominences through the Chromosphere filter and surface details through the Prominence unit.

The Quark is a 4.2× telecentric Barlow lens that incorporates a 21-millimeter Hα filter and a 12 millimeters blocking filter. All optical elements receive anti-reflective coatings at the factory. To further reduce scattered light (which decreases image contrast), the company has built the unit with a series of internal baffles. The Quark tips the scales at 13.9 ounces (394 grams).

Hα filters that attach to the eyepiece end of a refractor often require an extra energy-rejection (ER) filter in front of the objective lens (that is, between it and the Sun). That pre-filter gets rid of much of the harmful radiation before it enters the main filter. If you mount the Quark on a 6-inch (150 millimeters) or larger scope, use an ER filter at the front of the telescope. The other time you'll need to use an ER filter is if your telescope has optics at other points within the tube. One example of this is the Petzval design. Without an ER filter, the two lens elements inside the tube would become too hot.

If your objective is less than 6 inches in diameter, you can use an optional UV/IR blocking filter Daystar sells (DSIUV2, $120). Just screw the filter onto the front of your star diagonal, and drop the Quark into the other end.

Fig. 16.7 One accessory for the Quark you'll want to consider is the UV/IR blocking filter. (Courtesy of *Astronomy* magazine)

Daystar states that you can use the Quark without any ER filter for occasional views of the Sun on telescopes with apertures less than 3.15 inches (80 millimeters) and non-tracking mounts. In both cases, excessive heat wouldn't build up within the scope. And, indeed, I had no problems viewing through my 3-inch alt-azimuth-mounted refractor without a filter off and on for several hours.

The only knob on the Quark controls the heater, which really determines what the central wavelength is. Each detent on the knob shifts it by 0.1 Å, and the filter allows a 0.5-Å adjustment either way. This puts different details "on band." Just be sure to give the filter a few minutes to attain the new temperature once you rotate the knob. One nice feature the company included is an LED that changes from yellow to green when the filter is on band.

Daystar claims that from a cold start the Quark takes approximately 15 minutes to reach operating temperature (the LED turns green). I can confirm that length of time as an average, although on one sunny 20° Fahrenheit (–7° Celsius) day in Wisconsin when I used the optional power supply, the heating element required an additional 5 minutes to warm up (and I never did get warm). Nobody will have this problem on eclipse day. Once the LED turns green, you'll see prominences, faculae, flares, and, yes, even sunspots.

One of the cool things about solar viewing is that some of the Sun's features, mainly prominences and flares, show noticeable changes in real time—in just minutes. I saw tree-shaped prominences reform into arcs and then to spouts within 15 minutes. And in twice that time, I observed flares, which appear as bright (even overexposed) regions, sending out tendrils from which new flares formed.

Three things make the Quark great fun to use. First, it's not a complicated setup: plug in power; insert into focuser; and observe. Second, the device is ultraportable. It and its power supply (bubble wrapped) fit into my fabric telescope case. And third, when coupled with my 3-inch refractor, views through the Quark surpassed my expectations.

I love observing the Sun and turning people on to how utterly cool it is. Daystar has given me something new to suggest to them that, at a price of $995, won't break the bank. You can get more information online at www.daystarfilters.com.

Daystar Quantum SE

Daystar makes more filters than just the Quark. Let's look at the company's 0.5 Å Quantum SE Hα filter. Daystar makes two types of Quantum filters. The SE series fills the needs of most amateur astronomers and educators. The PE series is designed for professional research applications.

The SE filter requires a heated etalon, an optical filter that has a narrow bandwidth at the required wavelength. Daystar produces Hα filter systems in 0.3 Å to 0.8 Å bandwidths in 0.1 Å increments. A measure of light wavelength, 1 Å equals one ten-billionth of a meter. Narrower bandwidth filters—say, from 0.3 to 0.5 Å filter—increase the contrast of solar images.

Fig. 16.8 Daystar Filters' 0.5-angström Quantum SE Hydrogen-alpha filter will allow you to observe and photograph the Sun's chromosphere, prominences, and flares. (Courtesy of *Astronomy* magazine)

Daystar incorporates a green light that indicates when the filter is on-band. Easy-to-use redshift and blueshift buttons tune the wavelength up or down precisely. A serial port allows computer control capability and filter readout display.

For the review, I also used a Daystar full-aperture energy rejection (ER) filter. The ER filter mounts on the front end of the telescope. It reduces the heat load on the SE's assembly by absorbing or reflecting most visible light. Reducing the heat load lets the Quantum filter function more efficiently. It also gives it a longer usable life. Reducing the amount of energy that strikes the filter helps protect the system's blocker and trimmer elements, which tend to degrade with use.

When the filter is mounted on various telescopes, you may have to use a Barlow lens to achieve an effective focal ratio around f/30. The system requires such a focal ratio because the light entering the filter must be as close to parallel as possible. Faster f/ratios (below f/30) allow light to strike the filter at steeper angles, providing a sub-par performance.

You can increase your optical system's focal ratio in two ways: either with a Barlow lens or by employing an aperture mask, which reduces the aperture size, thus increasing the focal ratio. Aperture masks also reduce the light throughput, but this is less a problem when you're dealing with the Sun.

Fig. 16.9 An energy-rejection filter attaches to the front of the telescope. It blocks ultraviolet, infrared, and some visible light, reducing the heat on the Hydrogen-alpha filter. (Courtesy of *Astronomy* magazine)

The Barlow-plus-filter system is easy to set up because the directions supplied with the filter are foolproof. The Quantum SE operates on 12 volts of DC power. Once it powered up, it took about 8 minutes to warm the optics to the proper temperature. As the filter heated up, it reached 6,562.8 Å and stopped. The indicator light turned from yellow to green. Daystar's digital readout clearly shows the current wavelength—a terrific feature. The readout has an accuracy of 0.1 Å. It can also display error codes, such as voltage variations or internal malfunctions.

You'll quickly determine what proper focus should look like and how to tune the filter using the red and blue buttons. Those buttons change the filter's wavelength by small amounts. This allows the observer to adjust the filter's temperature for the best image. Some Hα filter systems tilt the etalon to make this change.

Based on my experience, a high-quality, short focal-length refractor produces excellent solar images. A longer-focal-length optical tube assembly would make achieving the f/30 recommended effective focal ratio easier, but a Barlow lens and/or an aperture mask can help you do that. With Daystar's Quantum SE Hα filter, even a quiet Sun—like what solar physicists predict for August 21, 2017—will look spectacular.

Coronado PST

In the early 1990s, Hα telescopes were rare and brutally expensive. Within a few years, however, new suppliers, improved filter designs, and Hα solar telescopes that were self-contained appeared on the market. Selection and performance improved and prices dropped, some. Hydrogen-alpha observing was still expensive, but more tolerable than before. It seemed amateur solar observing might be entering a golden age. Still, nobody was prepared for an announcement that a complete Hα solar telescope soon would appear and cost only $499 including the eyepiece.

Fig. 16.10 Coronado Instruments' Personal Solar Telescope (PST) is ready to attach to a camera tripod right out of its case. The MALTA tabletop mount, shown here, is optional. (Courtesy of *Astronomy* magazine)

Coronado Technology Group's Personal Solar Telescope (PST) is a self-contained Hα solar-observing rig. The PST is a 1.6-inch refractor with an integral Hα filter. It accepts 1¼″ eyepieces; the manufacturer provides a 12.5 millimeters (32×) Kellner eyepiece. The instrument measures 15 by 2.1 by 3 inches (38.1 by 5.3 by 7.6 centimeters) and weighs 3 pounds (1.4 kilograms). It has a small knob to control focusing, which is done internally, and sports a ring at the base of the brass tube that allows the user to tune the image. Tuning allows for better views of certain parts of the solar disk. Tune the PST's filter slightly, and the chromosphere is more prominent; tune it a bit differently, and prominences will stand out.

There is also a solar finder, which the maker calls a "Sol Ranger," built into the telescope's body. This is one of the PST's delights. When you point the PST in the general direction of the Sun, a small image of the Sun appears in a circular window of semitransparent material near the eyepiece—a totally safe view.

When you center that image, the Sun is in the telescope's field of view. Frankly, the little finder is amazing. It contains no projecting or fragile parts, uses no power, never needs calibration, and works accurately and reliably. The PST offers crisp, clear images with no sign of ghosting (a faint copy or copies of the main image) and no significant scattered light in the field of view. The Sun appeared bright, but not so bright that features looked washed out.

All the basic features one would expect to see through an Hα filter were visible through the PST. You can watch prominences grow and dissipate in real time. Active regions generally show the underlying sunspot if it was large, but the contrast on the disk of the Sun is still sufficient to watch fine movement of the solar atmosphere caused by shifting magnetic fields.

If you want to use your own eyepieces, the PST supports magnifications up to 100× before the image becomes excessively fuzzy. During periods of good seeing, try 70× to reveal details you won't perceive easily at lower magnifications.

Fig. 16.11 The black ring allows users of Coronado's PST to "tune" the image. A gentle turn reveals solar prominences; turn the ring the other way, and you'll see chromosphere detail. (Courtesy of *Astronomy* magazine)

You can add the optional MALTA tabletop mount, a convenient but expensive accessory that travels in the same case as the telescope. The MALTA has smooth motions and works pretty well—this accessory is not a gimmick. It's the perfect height for setting the PST on the trunk of a car or a picnic table, but its short legs are not suited for deployment directly on the ground. Most observers prefer mounting the PST to a photo tripod. This arrangement offers a bit less stability than the MALTA but puts the telescope at a more convenient height and offers a bit smoother motion. Attaching the PST to a tripod is easy, thanks to a ¼-20 thread mounting hole. With either mount, setting up the PST takes less than a minute, which hopefully will encourage you to look at the Sun frequently.

As you'd expect, the PST's image is not as contrasty and doesn't show as many details as larger Hα scopes. While the PST shows prominences just as readily, it doesn't offer as much surface detail. Its principal compromise is price vs. performance. It performs well, but not spectacularly. The images are good, but fall somewhere short of great, especially if you have something better to compare them with.

On the other hand, the price is low: For better images, you would pay a lot more. The PST also is well built and rugged. Normal use and even a bit of unintentional abuse had no effect on it. But perhaps the PST's strongest point is that it is undeniably convenient to use. It throws no obstacles in the way of the observer and encourages frequent observing. And that's a big plus. Learn more about Coronado's Personal Solar Telescope at www.meade.com/products/coronado.html.

Coronado SolarMax II 60

SolarMax II 60 Telescope is another instrument made by Coronado. This is a much pricier option than the PST, so I'm going to leave in more of the tech speak because mostly advanced amateur astronomers will consider this option. The SolarMax 60 was part of a line of Hα telescopes and filters introduced in August 2010.

Fig. 16.12 Coronado Instruments' SolarMax II 60 is a Hydrogen-alpha telescope. Unlike a visible-light solar filter, which can show sunspots, the SolarMax II's filter reveals prominences, flares, and the Sun's chromosphere. (Courtesy of *Astronomy* magazine)

This refractor features 2.4 inches (60 millimeters) of aperture, a focal length of 400 millimeters, and a focal ratio of f/6.7. Its filter is a two-piece, full-aperture etalon—an optical interferometer that bounces light between two partially reflective mirrors. The etalon sits in front of the 60 millimeters objective. The central wavelength of the light it transmits is the Hα line, and the width of the transmitted light is 0.7 Å (1 angstrom=0.1 nanometer).

With this scope, Coronado also introduced its RichView system, which works by letting you tune the etalon. Tuning allows you to slightly adjust the central wavelength of the transmitted light. Altering it slightly one way or the other provides for the right combination of viewing either prominences or features in the solar chromosphere.

When the telescope is not in use, a threaded metal cover protects the objective and the etalon. The SolarMax II system includes a blocking (energy-rejection) filter, the 0.7 Å etalon, a diagonal, a 25 millimeters Coronado Cemax eyepiece specifically designed for use with the telescope, mounting rings, and a Sol Ranger solar finder scope.

The SolarMax II also comes with a hard carry case as suitable for travel as it is for storage. The instrument sports a beautiful brass tube and black finish. A dovetail mounting plate connected to two clamshell rings makes it simple to attach the scope to a mount. Attach the diagonal and the 25 millimeters eyepiece, and you'll be ready to observe the Sun. The drawtube moves back and forth in the optical tube, which allows for easy rough focusing. I then used the telescope's helical focuser for fine focusing. If you're like me, your first reaction will be "What a view!"

Fig. 16.13 The lever to the right controls the tunable etalon in the SolarMax II. It tilts the front filter stack and thus moves the center of the 0.7-angström transmission band. One setting will provide a good view of the Sun's prominences, while another will show surface details better. (Courtesy of *Astronomy* magazine)

Surface details are easy to discern, including prominences along the Sun's edge. Those features appeared structured, especially with slight focus adjustments and by tuning the telescope's etalon via the SolarMax II's RichView lever. Feel free to substitute your eyepieces instead of the one Coronado supplies. A friend uses a zoom eyepiece that varies its focal length from 25 to 7 millimeters. He said that the ability to zoom in and out on solar features without changing eyepieces sure saves time.

The Coronado SolarMax II 60 Telescope is available in a number of optional configurations. You can get 5, 10, and 15 millimeters blocking filters (for telescopes of increasing focal lengths) and even a double-stack etalon. That option reduces the normal bandwidth of the telescope from 0.7 Å to 0.5 Å, which increases the visibility of certain solar features.

Frankly, solar viewing through a "white-light" filter gives an observer just a peek at the Sun's majesty. But pick an Hα scope, and you might find yourself at the eyepiece for hours on every clear day. The Coronado SolarMax II 60 Telescope is a superb choice if you want to upgrade to a dramatic view of the Sun. You'll find all the options for this instrument at www.meade.com/products/coronado.html.

Chapter 17

25 Tips
for Photographing
the Eclipse

Despite my earlier advisories to savor the experience of the eclipse rather than trying to take a photo of it, many people simply will ignore my sage advice and ruin their viewing experience trying to snap pictures. If you happen to be one of them, this chapter will try to help you as much as possible. While photography is *not* recommended as an activity during the eclipse, avid photographers can be hard to dissuade.

1. Select Your Equipment Well in Advance

You need to decide soon what optics you'll be shooting through. It could be a removable camera lens, a non-removable lens on a point-and-shoot camera, or the optical tube of a telescope. For telescope recommendations, see Chapter 14. For camera recommendations, please head to Chapter 15. You also must decide what you'll be shooting with. This boils down to three choices for static imaging: a digital single-lens reflex camera (DSLR), a point-and-shoot camera, or your cellphone. There are also a variety of choices if you want to capture video of the event. Oh, and you probably already know that you'll need an approved solar filter for your optics as a first priority!

© Springer International Publishing Switzerland 2016   181
M.E. Bakich, *Your Guide to the 2017 Total Solar Eclipse*, The Patrick Moore
Practical Astronomy Series, DOI 10.1007/978-3-319-27632-8_17

2. Practice with Your Equipment Well in Advance

Planning on taking your first pictures of the Sun on eclipse day is no plan at all. I'm not suggesting you head out every day from now until August 21, 2017, and photograph the Sun. But practice will pay off. You won't be as likely to fumble during the eclipse — and, believe me, that day will have enough tension and drama without you adding to it by making mistakes.

As they say, an amateur is someone who practices until they get it right. A professional is someone who practices *until they can't get it wrong.* While you don't have to drill your photo skills until you're a pro, you do need to at least reach the amateur level to have success on eclipse day. Now that you know that you have to practice, how to do that exactly? Simply go outside on a sunny day, aim your filtered camera at the Sun, and shoot. Any days of practice are helpful, but for the absolute best choices of a dry run, see Chapter 9. During your practice run, take a range of different exposures. This is the digital age, when no one has to pay for film or development. It doesn't cost anything to shoot 100 images, so grab a shot for all your settings. You don't even have to keep track of them because today's cameras record all relevant details about each image you shoot. They store this information in an EXIF file. Lots of web photo apps like Flickr show this data, and you also can see it if you use Photoshop.

After shooting, examine all the images. Toss the ones you don't like and narrow down to the best two or three. That's when to look at the EXIF file. In addition to the date and time of the exposure, you'll find the aperture (f-stop), focal length, ISO setting, exposure time, and more.

If you're shooting in manual mode, you really won't have to worry about exposure times until the eclipse nears totality. The Sun's surface brightness remains pretty constant throughout the eclipse, so you won't have to change exposure times until the Sun's disk narrows into a thin crescent. At that point, you can add two more stops due to solar limb darkening. Bracketing by several stops is also necessary if haze or clouds interfere on eclipse day.

3. Decide What You Want to Shoot

Page through the images in this book. Look through back issues of *Astronomy* magazine, or check out archived "Picture of the Day" images on the Astronomy. com website. Head to Google Images. Check out some of the URLs in Appendix A. In other words, see how previous eclipse photographers have captured total solar eclipses and decide if you want to try to replicate one of their shots.

Fig. 17.1 Eclipse sequences make some of the most spectacular images of the event. Most imagers shoot individual frames and later combine them with processing software like Photoshop. (Courtesy of Ben Cooper/LaunchPhotography.com)

Some imagers mount their cameras on a tripod and take single wide-angle shots during totality. These capture the Sun's corona and part of the earthly scene below.

Others take close-up shots of the Sun, either of the entire eclipse with all the partial phases or just the few minutes of totality. Most top-notch eclipse photographers who go this route will mount their cameras on motorized telescope mounts, which track the Sun's position as it moves across the sky. That way, they're not constantly having to alter their camera's position atop the tripod and re-center the Sun by looking through the viewfinder or at the camera's LCD screen.

Still others take a sequence of wide-angle shots that show the progression of the eclipse. For this, they aim the camera at the Sun's position at mid-eclipse, lock it into place, and shoot at equal intervals such as once every 5 minutes. Later, these shots can be arranged into a montage, or the user can combine them into one picture with image-processing software like Photoshop.

Fig. 17.2 Well before the 2010 total solar eclipse, residents of Easter Island set up to sell food and memorabilia to eager eclipse watchers. Pictures like this help capture what else was happening. (Courtesy of Holley Y. Bakich)

4. Arrive and Set Up Hours Before the Eclipse Starts

After you've done your practice solar photography, the day of the eclipse will finally dawn. To make the most of it, you can't be scrambling to set up in the moments before the big event. Instead, stake out a location for doing your eclipse photography well ahead. The advantages to setting up in advance on the day of are numerous. You can pick out a prime spot where you'll be able to set your equipment up, test it in advance, and deal with any problems. Plus, the extra time at the site will allow you to interact with other eclipse-watchers before you lose yourself in photography. You can even bring a picnic!

5. Photograph Everything

If you have access to a second camera, don't forget to chronicle what's happening around you as the event approaches. These can be the best photos to take; they don't interfere with your experience of the eclipse, and can document the activities of your astronomy club, friends, or family. These photographic mementoes are better ones to focus on, since they can't be replicated by another observer (or outclassed by a higher resolution camera).

Fig. 17.3 A high-capacity memory card like this one is cheap. Get several before the eclipse. (Photo courtesy of the author)

6. Bring Extra Batteries

Put extra batteries on your checklist now to avoid a dreaded shutdown of your equipment at the crucial moment. For every piece of equipment that takes batteries, pack extras to have on-hand—and make sure any devices that need to charge have been fully charged in advance. The best equipment in the world is of no use if it doesn't have power!

7. Check Your Camera's Memory Chip

On a similar note, be absolutely sure that the memory card in your camera has at least twice the amount of free memory you think you'll need to adequately photograph the eclipse. In fact, this event is rare enough that I suggest you purchase an additional memory card with at least 32 gigabytes of memory. The cost for a new SanDisk 32Gb card as I write this was $10 on eBay, and $21 for a 64Gb card. These prices probably will be lower by the time you read this.

8. Prep During First Contact

At Rosecrans Memorial Airport in St. Joseph, Missouri, first contact (the moment the Moon's disk first touches the Sun) through fourth contact (the moment the Moon's disk last touches the Sun) lasts 2 hours 53 minutes and 53 seconds. The entire eclipse lasts just under 3 hours at *all* locations throughout the U.S., but almost half of that time is a slow build-up. Remember, this event is *all about totality*. If you're concentrating on capturing those two and a half minutes, the hour after first contact is your final check to see if everything is working. You can still be social, but rehearse your plan at least a few more times. Then, about 25 minutes before totality, every great eclipse photographer I know moves into "image mode."

9. Secure Your Filter

Most solar filters made for telescopes fit tightly on the front of the tube. Camera lenses, however, come in a much wider variety of sizes. If your filter is even slightly loose, secure it with blue painter's tape. That product, available at any home improvement store, will not leave a residue on your equipment. While it's true that securing your filter is vital, so is the ability to easily remove it. I remember one would-be photographer during the eclipse expedition I led to Peru for the November 3, 1994, total solar eclipse who had such a problem removing his filter from his camera lens that he ruined his careful alignment and never was able to return the eclipsed Sun to the field of view.

10. Totality Lasts 8 Seconds

OK, not really. But I often quote an old friend of mine, Norm Sperling, who wrote a "Forum" in the August 1980 issue of *Astronomy* magazine titled, "Sperling's Eight-Second Law." I'll just reproduce the beginning here.

"Everyone who sees a total solar eclipse remembers it forever. It overwhelms the senses, and the soul as well—the curdling doom of the onrushing umbra, the otherworldly pink prominences, and the ethereal pearly corona. And incredibly soon, totality terminates.

"Then it hits you: 'It was supposed to last a few minutes—but that couldn't have been true. It only seemed to last eight seconds!'"

Get the point? If you encounter a problem—any problem—that takes more than a few seconds, stop! Forget about photographing the eclipse, and gaze in wonder at what's going on above you. You'll have another chance (under an even longer duration of totality) in April 2024, the next time the Moon's dark inner shadow touches the continental U.S. By the way, if you'd like to read Sperling's entire column, jump online and head to www.fpsci.com/Sperling8seconds.pdf.

11. Calculate the Field of View (FOV) for Your System

I probably could insert a table here that would take up 20 pages. So many DSLRs exist today, and they all seem to have a variety of lenses they can accept. Plus, not all of them have the same size sensors (chips), and sensor size factors mightily into the field of view calculation. So, instead of an unwieldy table, here's an easy way to figure out a lens' field of view *on your camera*.

$$FOV = 2 * \arctan(0.5 * s / f) * 180 /$$

FOV = Field of view, s = Sensor dimension in millimeters, and f = Focal length of the lens in millimeters.

Do note that this formula is for just one dimension (x or y; that is width or height) of your chip. If you want to figure out the field of view in both dimensions, you'll have to perform this calculation twice. An example probably will help.

Let's say your camera is a Canon 6D and you want to know what the field of view would be if you choose to photograph through a 200 millimeters lens. Camera buffs call the 6D's sensor (chip) either "full frame" or "35 millimeters." That means it measures the same as a piece of film in an old 35 millimeters camera, or 36 by 24 millimeters. (NOTE: For exactness, the Canon 6D chip actually measures 35.8 by 23.9 millimeters, but the round-off is accurate enough for our purposes here.) The vast majority of cameras have smaller sensors.

The horizontal field of view would be $2 * \arctan(0.5 * 36/200) * 57.3$, or 10.3°. If your calculator doesn't have an arctangent key (also labeled tan-1), you can find numerous scientific calculators online. Now let's do the vertical field. While you could use the formula again, why not make things easy? The vertical measures $24/36 = 0.6667$ of the horizontal, or $10.3° * 0.6667 = 6.9°$. Therefore your field of view for just this lens attached to just this camera measures 10.3° by 6.9°.

12. A telescope's Field of View

But what if you'll be attaching your Canon 6D body to a telescope rather than to a camera lens to photograph during the time starting just before and ending just after totality? Believe it or not, the calculation is the same. In this case, however, your telescope substitutes for a camera lens. So, you can still use the above formula; just insert your scope's focal length, and you'll have your answer. If you're new to telescopes, most manufacturers print the focal length of the telescope either directly on the tube or on the ring that secures the front optic. Not in either place? Just look in your instruction manual.

13. Get the Right Camera-to-Scope Adapter

This relates to the previous item. Technically, you'll be coupling your camera's body to the telescope's focuser, but you can't do that without a two-part adapter. You'll need both a T-ring and a T adapter. The T-ring, like one of your lenses, has the male extensions that fit into your camera body's female slots. Turn it, and it will lock in place. Release it the same way you release a lens.

Fig. 17.4 A T-ring (*left*) attaches to your camera like a lens does. A T-adapter screws into the T-ring. The smaller diameter chromed barrel goes into the telescope's focuser. (Photo courtesy of the author)

The T adapter screws into the T-ring (most celestial photographers just leave the two parts connected). The adapter's other side is a 1¼"-diameter tube that slides into your telescope's focuser just like a 1¼" eyepiece. If your telescope has a 2" focuser, you'll need a 2"-to-1¼" adapter, but most manufacturers who use the larger focuser also provide this adapter.

14. Use a Remote Shutter Release

Why take the chance that you touching your camera will move it in a way you don't intend? Devices that trip your camera's shutter are small, easy to use, wireless, and inexpensive.

15. Calculate the Sun's Size on Your Chip

After you figure out the field of view of your camera/lens combination, it's a simple matter to calculate how much of your camera's sensor the Sun will cover. In the example in #11 above, the width of the field of view is 10.3°. The Sun and Moon both have an angular diameter of 0.5°. So the Sun's width on your image will be $0.5/10.3 = 4.9$ percent (call it 5 percent) of the field.

16. Calculate the Corona's Size on Your Chip

If you're photographing totality, the Sun's disk is just the beginning. In fact, between second and third contacts you won't see its disk at all. What will still be visible, however, is the evanescent corona. But how far from the Sun's limb (edge) will it stretch? My best guess, which agrees with the half dozen or so eclipse experts I asked, is between two and three solar radii. If we stick with the 6D and a 200 millimeters lens, a corona measuring two solar radii will have a diameter of 2.5°. So, $2.5/10.3 = 24$ percent, or one-fourth as wide as the field of view (and 36 percent as high). A three-solar-radii corona will span 34 percent, or one-third of the width (and 51 percent of the height).

17. Don't Go Nuts on Your Lens' Focal Length

This item highlights the two points that precede it. If you want to capture the entire Sun to document the partial phases, perhaps because you are unable to make it to the path of totality but still want to photograph the partial eclipse above your site, make sure you don't pick a lens or telescope that restricts the view to less than 0.5° in the vertical dimension. Increase this to 3°, or even a bit more, if you aim to photograph most of the corona. Remember, your camera will capture more than your eye can discern.

18. Avoid Small f-Ratios

If you're using a standard zoom lens or have added a teleconverter to your system, the image will not be sharp if you choose to shoot at f/1.4. Instead, pick an f/ratio from f/8 to f/11, and you'll get the sharpest images you can. Fortunately, we're dealing with the Sun, so there's plenty of light to go around even through such reduced apertures.

19. Focus Is Critical

Show someone an out-of-focus image you took of the eclipse, and they may not say anything about it. But I guarantee you they'll think, "Oh, wow, you spent all that time and money getting ready, and *this* is what you have to show for it?" Of course, people think similar thoughts about unfocused images of the Mona Lisa, the Hoover Dam, or Grandma Irene. We're used to seeing sharp pictures. I view this as a major reason why practicing beforehand is so important.

You may not know this, but most of today's camera lenses have the ability to focus past infinity. Back in the day, lenses didn't do that. Photographers would crank them all the way clockwise or counterclockwise and, boom! Perfect focus. Now you have to line up an arrow with the "sideways 8" on the lens. In my mind, this is not progress, but we have to live with it.

The easiest way to focus is to aim at a distant earthly object. Then don't touch the focus ring again. If such an instruction makes you paranoid, you can tape the focuser using the same painter's tape I mentioned in #9. Oh, and turn off your lens' autofocus. Pointing it at the sky wreaks havoc with the high-tech focusers manu-facturers use today.

20. Get Acquainted with Your Write Time

How fast you can take pictures depends on your camera-to-memory-card write time. The latest cameras have microprocessors and internal memories that have reduced this interval to next to nothing. Older digital models and many point-and-shoot cameras, however, are quite another story. Be familiar with how yours performs lest it delay your shots and surprise you by reducing the number of images you can capture.

21. Consider an Intervalometer

An intervalometer is a totally cool device that lets you take time-lapse photos. If your plan is to capture close-up images during the entire eclipse, a properly set intervalometer will let you take exposures every X minutes, where X is any interval you choose. And on a driven mount, you can capture images without having to fuss over your equipment. That means that you also can watch the event. Well, most of it, since I'm sure you'll still take a look at your setup to see that everything is proceeding as planned.

Search online, and you'll find a variety of intervalometers, also called timer remote controls, for many Canon and Nikon DSLRs.

Fig. 17.5 Eclipse photography becomes much easier and "hands-off" if you employ a timer remote controller. (Courtesy of Mike Reynolds)

22. No Flash

If you're using a point-and-shoot camera, be sure to turn off the flash. Leaving it on not only won't do any good, it will drain the camera's battery and may even annoy the people near you. In fact, if you see someone who will be using such a camera, you may want to share this point with him or her. In truth, the flash is just an annoyance. And although you'll surely hear someone say that a flash's light will ruin your dark adaption, even that's not true. Even during the darkest eclipse, the eye's dark adaption doesn't trigger during the few brief minutes of totality. And this eclipse won't be close to the darkest possible.

23. Use an Approved Solar Filter

As should be clear from how often it has been repeated, viewing and photographing the eclipse must only be done through an approved solar filter. For your eyes to be safe and comfortable visually, a filter must reduce the visible light by a factor of 160,000. Camera chips are a bit more forgiving, and the unit's ability to adjust exposure times pretty much guarantees success. So, attach a filter to your camera lens. Note that although a #14 welder's filter is approved for viewing the Sun at any time, it uses heavily tinted, one-quarter-inch-thick glass. This means that while fine for visual use, no one should shoot through these. You will be disappointed by the result.

Fig. 17.6 Before you shoot the partial phases, make sure you attach a solar filter to the front of your telescope—or telescopes! (Photo courtesy of the author)

24. Take the Filter Off to Capture the Corona

The Sun's disk or photosphere outshines the corona by 1 million to 1. That's why we never see the corona except during totality—even a smidgen of our star's visible surface completely overwhelms it. To observe or photograph the corona in all its glory, you must remove your solar viewing glasses, or put down your handheld filter, during totality.

One thing to keep in mind when you're thinking about capturing the corona is that its brightness varies according to its distance from the Sun. It's really bright next to the solar limb or edge, and it gets progressively fainter as you increase the distance to the limb. Usually, by the time the distance from the limb equals approximately one solar diameter, human eyes can no longer see the corona. But cameras can. Every eclipse photographer I know takes the time to bracket his or her exposures during totality. That way some will reveal detail in the inner corona while others show features in the much less dense outer corona. Enterprising individuals even have invented processing techniques that let them combine shots to show details in the entire corona.

25. Use a Sturdy Tripod

Absolutely none of the photography I've described in this chapter falls under the "handheld" variety. Shoot away with your cellphones and point-and-shoot cameras if you want, but honestly, you are wasting your time and throwing away seconds when you could be awe-struck by the celestial spectacle in the sky above you. While image stabilization sounds good incorporated into upscale lenses, nobody experienced trusts it for capturing eclipses.

You have two choices for eclipse photography. First, you can use a standard tripod. Or, you can attach your camera to a telescope mount (with or without the telescope) that sits on a tripod.

If you're worried about the sturdiness of your tripod, you can help matters out by hanging a 10–20 pound weight centered under your tripod's head. Doing this will lower the unit's center of gravity, making it more stable.

Finally, discourage people from approaching your photo setup. Why take the chance that Aunt Mary could accidently kick one of the legs of your tripod? That would ruin your alignment and immediately turn you into a visual observer. Hmm, on second thought, would that be so bad?

And here's a bonus.

Process and Share Your Images Quickly

Can you imagine how many people will be photographing the eclipse? If you want your images to stand out or even to be published somewhere, whether in *Astronomy* or elsewhere, you have to send them out soon after the eclipse is over, certainly no later than the end of the day August 21. An email address that will surely be working is ReaderGallery@Astronomy.com.

Chapter 18

Projects
for Observing the Sun
(and the Eclipse)

How to Build a Simple Sun Viewer

Let's say you want to view the Sun, but you don't have an approved solar filter for your telescope. Maybe you don't even have a telescope. No problem. Somewhere at home you must have a cardboard box. That and a few common supplies will let you build a pinhole viewer that you can observe the Sun with. While I admit that the image you'll see won't rival that through a properly filtered telescope, it will be the Sun, and you can watch it as long as you want without any safety concerns. Plus, building this viewer is essentially free.

Just follow the steps in the captions. Feel free to experiment with the size of the box, whether or not you leave the extra cardboard pieces attached, or the pin's diameter. In fact, if you're a clever soul, you may want to mount the box on a thin plywood base, which you then can attach to a sturdy camera tripod. That way, you won't have to hold the box while viewing the Sun. This activity is great for school classes. In that case, I suggest the teacher cuts out the hole for an additional safety factor.

© Springer International Publishing Switzerland 2016

M.E. Bakich, *Your Guide to the 2017 Total Solar Eclipse*, The Patrick Moore Practical Astronomy Series, DOI 10.1007/978-3-319-27632-8_18

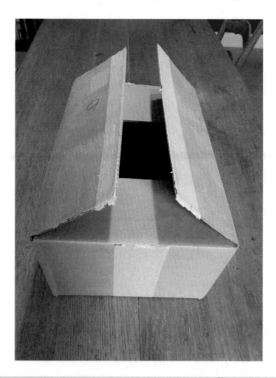

Fig. 18.1 Step 1: Start with a cardboard box at least 18 inches long. If you use a box shorter than this, the projected image of the Sun will be unacceptably small. (Photos in this chapter courtesy of the author)

Fig. 18.2 Step 2: On one of the box's smaller ends, trace a circle by using a quarter

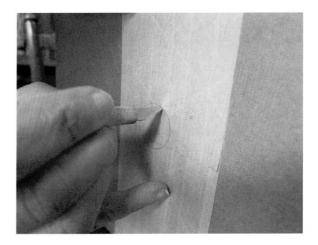

Fig. 18.3 Step 3: Carefully cut out the circle with a sharp knife

Fig. 18.4 Step 4: Note that the hole doesn't have to be perfect—or even round!

Fig. 18.5 Step 5: Cut out a small piece of aluminum foil big enough to cover the hole in the box. The best foil to use is a fairly thick variety

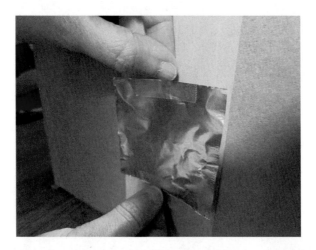

Fig. 18.6 Step 6: Tape the foil over the hole

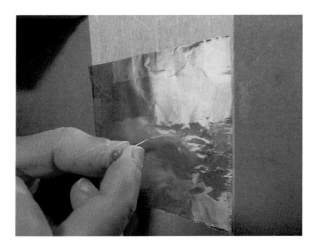

Fig. 18.7 Step 7: Poke a hole in the foil using a pin. A pin is better than most items you could use because you want a hole with clean (not ragged) edges. As a variation, you can try using a sharp pencil. That will produce a larger hole. Larger holes let more light through and so produce a brighter Sun image. Careful, though. If your hole is too big, you'll lose the pinhole camera effect and you'll just see an out-of-focus blob

Fig. 18.8 Step 8: Inspect the hole for roundness and make sure nothing is blocking it

Fig. 18.9 Step 9: Cut away as much of the extra cardboard pieces as you want. Alternatively, you could fold them over and tape them to the box's sides, making the assembly a bit sturdier

Fig. 18.10 Step 10: Tape a white piece of paper on the inside of the box opposite the hole

Fig. 18.11 Step 11: Point the hole at the Sun and observe the projection on the paper. Congratulations!

Project #2

Build a Solar Projector

Building a simple Sun viewer a bit too simple for you? Is the image quality not what you expected? Then try this project. I guarantee you'll like the results. You'll need a little lumber, a bit of hardware, and a telescope's finder scope that's optically configured for straight-through viewing. If you're careful, you'll be able to swap the finder scope back and forth from the solar projector you build to the telescope.

Actually, you might want to go one step farther. Buy a bit of sandpaper (a variety pack made of several different grits) so you can give whatever wood you choose a smooth finish, because nothing hurts like a splinter.

It doesn't really matter what type of finder scope you use. They range from extremely simple to "simple but more expensive." The larger the finder scope's front lens, the more it generally costs. Now it may sound funny, but it's lucky most manufacturers of finder scopes use fairly inexpensive eyepieces. Lucky because a complex, expensive multi-element eyepiece will trap more heat between its elements. Simple lenses will warm some but won't retain most of the longer-wavelength infrared radiation (heat).

The first measurement you'll need to make is the distance from the finder scope's eyepiece to where the Sun comes to a focus. It's really nice to have a second person's help for this. Go out and prop up a white piece of cardboard so it sits facing the Sun. You also can tape any white piece of paper to a board. Then hold the finder scope so it projects the Sun's image onto the cardboard or paper. It'll take a bit of finagling to hold steady, but it's not difficult. When you have a reasonably steady image, measure—or, better yet, have person number 2 measure—the distance between the eyepiece and the paper. Let's call that distance T, and here we'll measure it in inches. Feel free to use centimeters, cubits, or whatever you like. While you're at it, measure the diameter of the Sun's projected image. We'll call that measurement D.

The long narrow board you will mount everything on, which I call the long arm, must be T inches long plus enough extra length to accommodate the finder bracket and the thickness of the square piece of wood you'll use to project the image onto. I call that piece the screen board.

For the finder scope that I chose, T equals approximately 29¾ inches (75.6 centimeters) and D equals 4¼ inches (10.8 centimeters). The extra length needed to attach the finder scope is 7 inches (17.8 centimeters). I chose a screen board ¾-inch (1.9 centimeters) thick, and I suggest you do the same because you'll need to attach it to the long arm with two screws. Thinner boards may crack. Also, because D equals 4¼ inches, I made the screen board a 6-by-6-inch (15.2 by 15.2 centimeters) square. With a tiny bit of extra length on each end, this made my long arm 38 inches (96.5 centimeter) long. I chose to make that piece 2¼ inches (5.7 centimeters) wide, and the board already was ¾ inch thick. I chose that width because it was just a bit more than I needed to accommodate the base of the finder scope. Any additional width would just add useless weight.

Next, sand the long arm to a smooth finish so there won't be any splinters. Then attach the finder scope base using two 1-inch-long bolts. Each bolt is long enough to go through the finder's base, but not long enough to stick out from the bottom of the long arm. Once complete, drill two holes wide enough to accommodate lock washers and nuts attached to the bolts and deep enough to hide them from protruding. Then attach the screen board to the long arm with two 1½-inch-long (3.8 centimeters) drywall screws. Make sure it's perpendicular to the long arm and parallel with the finder scope's front lens.

The next step is drilling the hole that the tripod's bolt will screw into, securing it to your tripod. Don't just measure to the center of the long arm and drill. Rather, attach the finder scope base, the finder scope, and the screen board to the long arm. Then, carefully balancing the long arm on your two index fingers held about 30 inches (76 centimeters) apart, slowly move your fingers toward each other. They will meet at the point where the long arm's weight balances. Mark the spot, drill a hole, and tap it for a ¼-20 thread. Specifications say you should use a 13/64-inch bit, but when I'm tapping wood I use a slightly smaller one, either 11/64-inch or 3/16-inch. After you attach a piece of white paper to the screen board you should be ready to rock. Choose a tripod sturdy enough to carry the weight and steady the arm, and attach your solar projector to it.

A couple of final notes. First, if you have lens caps for your finder, use them. They will keep dirt off the optics. No caps? Use two small kitchen sandwich bags secured with two rubber bands. Second, when you're transporting the solar projector, disconnect the finder scope and put it somewhere it won't get abused. Third, don't over-tighten the tripod-mounting bolt. Remember, you're screwing it into wood. If you do strip the threads, home supply stores sell metal ¼-20 inserts you can attach (usually, you hammer them into place), which you can screw the tripod's bolt into.

Fig. 18.12 To make the solar projector, I used an 8×50 finder scope that was lying around

Fig. 18.13 I mounted the finder scope bracket on one side of the long arm, which I made out of cherry and sanded to a smooth finish. No splinters here

Fig. 18.14 The screen board attaches to the end of the long arm opposite the finder scope. The board is ¾ inch thick and attaches to the board with two screws

Fig. 18.15 The holes beneath the finder scope's mounting bracket are large enough to accom-modate the screw and nuts without them protruding beyond the bottom of the long arm

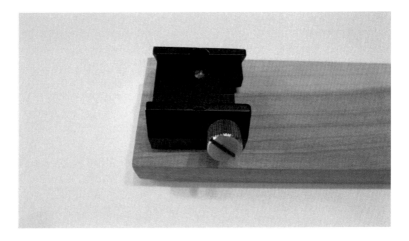

Fig. 18.16 The finder scope used for this solar projector focuses when you loosen the knurled knob, move (by rotating) the front lens some, and then retighten the knob

Fig. 18.17 The completed solar projector is simple, disassembles easily, and works like a charm. Note that the tripod-mounting hole is nowhere near the center of the board's long dimension. That's because the finder scope and its base weigh significantly more than the screen board

Project #3

Build a Solar Projector from Binoculars

If you're not an amateur astronomer, you may not have a finder scope lying around to construct Project #2. It's likely, however, that you have binoculars, possibly some that you haven't used in years. Here's how to make the same kind of viewer using them.

For this project, I personally chose not to use any of the two binoculars I own. Each is high-quality and expensive, and I didn't relish the thought of sending unfiltered sunlight through the multiple optical elements within each unit. Instead, I ventured online and spent $24 for the 7×35 binoculars pictured in this project.

They have a much less complex optical path than the ones I own. And you know what? They work great!

Something to remember if you try this project is that most of you will be mounting the finished product on a camera tripod. So try to save weight wherever possible. I strove to make this unit as small as practical and still provide the support the binoculars needed. I also performed a totally unnecessary step. I drilled a number of 1-inch diameter holes along the base's length. This will reduce the unit's weight without sacrificing much strength, though the amount of weight probably is insignificant.

The parts that make this projector are a wooden base, a metal L bracket, a metal plate the tripod bolt can screw into, and a wooden support for a screen. The binoculars I used measure 7¼ by 5¼ inches, but you don't need to make the base 5¼ inches wide. It's only there to provide attachment points for the L bracket and the camera tripod.

My base is 2 inches wide and 26 inches long. I estimated its length by letting the Sun's light pass through the binoculars and measuring how far from the eyepiece the image focused. For this unit, that focus length was 20 inches. That's the distance I determined the Sun's image was large enough to easily see, but not so large that it became faint. With this in mind, the base had to be a bit more than 5¼ plus 20 inches long.

I mounted the L bracket to the base so the ends of the binocular's front lenses were even with the front end of the base. I didn't want any of the base sticking out past the binoculars. I used wood screws to secure the L bracket to the base and a short ¼-20 bolt to attach the binoculars to the bracket. If you use a lock washer with the bolt, the bolt will be less likely to loosen, so your binoculars will be less likely to shift.

The only consideration for the L bracket is that it holds the binoculars high enough so the Sun's image isn't cut off at the base. The easiest way to figure out the height you'll need is first to measure the diameter of the Sun's focused image (that you created two paragraphs ago). Then take one-half that length and add ½ inch to it.

I then attached the 8 inches by 8 inches wooden piece (I used ¼ inch-thick plywood) that would serve as the backstop to a white piece of thin cardboard onto which the binoculars will project the Sun's image. A couple of small wood screws will secure it forever.

The next step is attaching the plate that the tripod's bolt will screw into. I used a 1 inch wide ¼ inch thick piece of steel. But you won't be attaching this at the middle of the base's length. Rather, attach the binoculars and the screen board to the base first. Then, carefully balancing the base on your two index fingers with one at each end of the base, slowly move your fingers toward each other. They will meet at the point where the base's weight (with everything attached) balances. Mark the spot, also centered according to the base's width. That's where to place the centered tripod-mounting hole in the 1 × ¼ inch bar. In fact, you should see your mark at the bottom of the hole. Remove the binoculars, L bracket, and screen board, and attach the metal bar.

If you're a bit unsteady or unsure about manipulating the base on your fingers with everything attached, there's another way you can find its center point. Make a fulcrum from a wooden dowel, a short piece of electrical conduit, or some other long round thing. Its diameter can be anything from ¼ inch to 2 inches. Then place the base with everything attached onto the fulcrum, move the base until you find where it balances, and mark it.

Once you assemble the unit and mount it on a tripod, it's your choice whether to use one side of the binoculars or both if the images don't overlap. If you're using just one side of the binoculars, the easiest way to eliminate the other side is to leave the rubber/plastic caps on it. I like the effect of having two Suns visible.

Fig. 18.18 The base of my binocular projector is a piece of cherry that measures 26 inches by 2 inches by ¾ inch

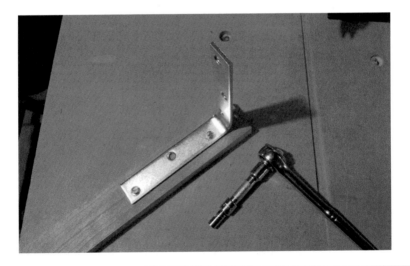

Fig. 18.19 I used two wood screws to attach the L bracket to the base. My L bracket measures 6 inches by 6 inches

Fig. 18.20 Nearly all binoculars have a ¼-20 threaded hole to allow you to attach a bracket

Fig. 18.21 The backstop is a piece of ¼ inch-thick plywood 8 inches on a side. I attached it to one of the base's ends with two wood screws

Fig. 18.22 With everything attached, I found the unit's balance point by slowly rolling it back and forth on a pencil

Fig. 18.23 The threaded hole of the tripod-mounting plate lies at the balance point of the binocular projector's long axis

Fig. 18.24 I use a white piece of paper or thin cardboard as a projection screen. Two spring clips hold it in place

Fig. 18.25 (**a, b**) To increase the contrast of the projected Suns, I cut a piece of foamcore board that fits over the front of the binoculars. Using this shield means the whole projection screen will be in shadow

Create a Fun "Sun Crescent" Sign

For this project, all you need is some white poster board and something to poke small holes in it. Lacking poster board, you also can use comic book backing boards. Each of these choices is thin enough to allow you to punch holes, but sturdy enough not to bend in mild breezes.

Do note that the cleaner your holes are the better your final image(s) will look. To that end, I purchased a cheap set of six leather punches online. They're all quite sharp and produce exceptionally clean holes. But which size to use?

Fig. 18.26 To test the size of the hole I wanted to use, I punched a series of various sizes into a piece of cardboard

This was easy enough to test. I made three holes with each of the six punches in the set. I then took the card I had punched outside with another white card and studied the Sun's images projected by it. I picked the one that looked the best to me, as should you.

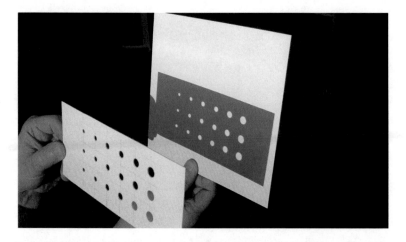

Fig. 18.27 I took my test card outside and let the projected images of the Sun help me pick the right size hole to use

After you perform these easy tasks, it's time to let your creative juices flow. Cut a piece of poster board roughly 7 inches by 3 inches. Then draw a pattern. Most people use a word, a name, a place, or the date of the eclipse. Mark dots often enough along your lines that you can recognize the word/date just by looking at the dots. Then, by using your punch, turn each dot into a hole.

Fig. 18.28 For your sign, draw a pattern meaningful to you. Make it as simple or as complex as you like. This one gives the eclipse's date

On eclipse day, as the passage of the Moon across the Sun's disk turns it into a crescent, project your word onto a slightly larger piece of poster board and have someone photograph you holding it. And here's another thought. As desperately as I want all of you to experience totality, I understand that some people will not be able to stand under the Moon's umbra. The previous four projects still will work, however. Each will show a partially eclipsed Sun, which, really, is also what they will show for people in the path of totality, too. That said, Project #4 will show the best results if the Moon's coverage of the Sun is 75 percent or greater.

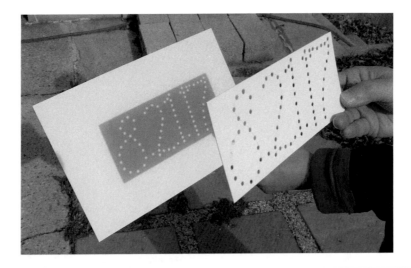

Fig. 18.29 The finished card lets you project multiple images of the Sun onto a second piece of white cardboard. And remember, on eclipse day you'll see a variety of phases projected. Take pictures throughout the partial phases, and pick the one you like best

Video Projects

Many of today's high-end digital cameras, and also not-so-high-end point-and-shoot cameras, include the option to record video. That opens them up to a wide array of uses during the August 21, 2017 total solar eclipse. The best thing (well, after the results you'll get) is that all you have to do is acquire some easy-to-find items and make some simple preparations. Then just hit the "Record" button, and walk away.

For projects #6, #7, and #9, in addition to the camera, you also will need a tripod. But if you plan to do any multiples of these projects, you don't necessarily need one tripod per camera (although you can choose to set up that way). If you follow my suggestion and build Project #5 first, you'll need only one tripod.

Camera Caddy

With the availability of inexpensive new as well as used point-and-shoot cameras, it's possible you may wish to carry out more than one of the following video projects. You even can invent your own. My thought was that I had invested enough in the cameras, and that I didn't want to buy multiple tripods. Other considerations come into play besides the cost, of course. Storage. Transporting them to the eclipse site. Carving out enough territory to set everything up. Ease of access.

To simplify my video projects, I came up with a simple device I made from materials lying around my shop coupled with an inexpensive online purchase. Here's how I did it.

Fig. 18.30 I started with a board 7½ inches square and ¾ inch thick. I used a piece of hardwood because I like the feel of it after sanding. Plywood, however, would work equally well

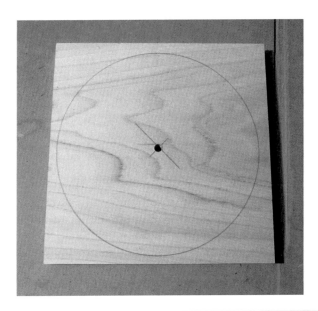

Fig. 18.31 I marked the center of the board and, using a compass, scribed a circle. I didn't actually measure the circle at the time, but it turned out to be approximately 6¾ inches in diameter. NOTE: You can leave the board square. I cut the edges off because I wanted to get rid of as much extra weight as possible and because I think it looks a bit more finished. Go with your preference. After drawing the circle, I drilled a ¼ inch hole in the board's center

Fig. 18.32 I then used a band saw to cut along the circle

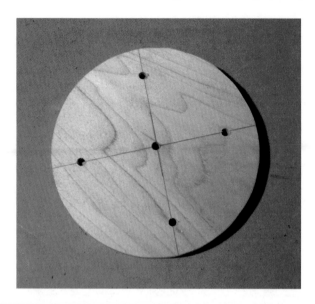

Fig. 18.33 After extending the initial centering marks, I drilled four ¼ inch holes, each 1 inch in from the edge

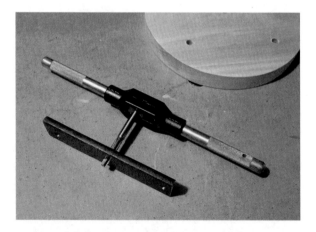

Fig. 18.34 I then found a piece of 1 inch wide and ¼ inch thick bar stock (you can purchase this at any home supply story). I cut mine about 4¾ inches long. I drilled and countersunk two mounting holes near each end and then drilled a hole that I tapped to ¼-20 size. Charts recommend a #7 drill for this purpose, or you can use the nearest fractional equivalent, 13/64 inches

Fig. 18.35 To hold the cameras, I purchased four (actually, I bought six) mini-ball-head hold-ers. Each of these measure only 2½ inches long and have a ¼-20 male end that screws into the bottoms of the cameras. I found them on eBay for $1.98 each, and shipping was free

Fig. 18.36 This picture shows the bottom of the Camera Caddy. I sanded each edge of the metal bar and the wooden base so I won't get an injury if I have to move fast on eclipse day

Fig. 18.37 And here's the top of the Camera Caddy with all four mini-ball-head mounts attached

Fig. 18.38 This last picture shows the completed project mounted atop a camera tripod

Record the Darkening

This is the simplest video project you can perform and as a bonus it requires no additional equipment besides a tripod. I have set this up for the past half dozen or so total solar eclipses I've experienced, and afterward I always was glad.

First, determine your "spot" for viewing the eclipse. It may be with your spouse, family, friends, or telescope, but you'll need to stake out an area before the eclipse begins.

Move your camera on its tripod far enough away from where you're set up so that it frames you completely. From this point onward, the camera will be facing directly away from the Sun's position at mid-eclipse.

At a specific time you've decided on in advance, begin recording video. I have used either 15 or 10 minutes prior to totality. Then just let the camera run until an equal amount of time has elapsed after totality, and stop recording.

You will then have a video record of the darkening (and subsequent re-brightening) that occurred during the most dramatic part of the eclipse. In addition, if your camera also records sound when in video mode (and pretty much every camera does), you will have captured your reactions, and probably those of people around you, to watch or share with others whenever the mood strikes you.

I have three important notes to add to these instructions. First, be sure your camera isn't set to compensate for decreasing light levels. You *want* the scene to darken. If it is, the model's instruction manual will reveal how to cancel it.

Second, make certain the chip in your camera has enough memory to record 20–30 minutes of video. Some cameras will display the number of "minutes left," on the card, or some such message, when you begin video recording. Others provide this information in the instruction manual, which will say something like, "If you use a 16Gb memory card, you can record XX minutes of video." If your camera does neither, then charge the battery, hit video record, and see how long it will go before your memory card fills up. Actually, if it's still going after half an hour, you can stop. That's plenty long enough.

Finally, make absolutely sure the battery you're using to record video is fully charged before you begin, and the spare battery too. This detail is easy to overlook.

Record the Shadow Passing Overhead

This project is just as simple as #6. First, however, there must be clouds at your location. That's not something I will wish on you. In fact, I hope your location has blue sky from horizon to horizon. If broken clouds (usually cumulus) exist, however, or if you can see distant clouds on the horizon, you may want to record the passage of the Moon's shadow as it speeds toward your location.

Simply point the camera in the direction of the approaching shadow, if clouds lie in that direction, or aim it in the direction the shadow travels after it covers your location if you see clouds there. Then follow the instructions for Project #6. For the August 21, 2017 eclipse, the Moon's shadow will race across the U.S. from the northwest (Oregon) to the southeast (South Carolina).

Project #8

Record the Temperature Drop

For this exercise, you'll need a digital thermometer and a digital watch or clock. Actually, an analog thermometer that contains mercury will work, but the temperature readings won't be as precise. Likewise, an analog watch or clock will work, but discerning the time won't be as quick. Please note: If you go the digital thermometer route, choose one that doesn't have a timer to shut its display off after a certain time; 10 minutes is the usual amount.

Fig. 18.39 You can record the temperature drop at your location by taking a video of a setup like this one. (Photo courtesy of the author)

One plus related to this project is that you don't need a tripod. You can set the three components on a table, a board, or, in a pinch, directly on the ground. If you have a way to shade the thermometer from direct sunlight, the temperature reading will be more accurate.

This last tip will apply to only some of you. If your camera has an intervalometer, an internal timer that trips the shutter at pre-set intervals, use it for this project. You really don't need a continuous reading of temperature because it won't change that fast. If, however, you can set your camera to record a frame every 1, 2, or 5 seconds, the movie you create will be 1/30, 1/60, or 1/150 times the size of a continuous one at 30 frames per second, and it will playback that much faster.

Project #9

Record Shadow Bands

I've observed a dozen totalities. In those, I've seen shadow bands once, and I wasn't even looking for them. That lone sighting occurred during the total solar eclipse November 13, 2012, that crossed northern Australia.

Shadow bands are undulating parallel lines sometimes seen in the few seconds before or after totality. The first person to describe them—but certainly not the first to observe them—was German-French astronomer Hermann Mayor Salomon Goldschmidt (1802–1866), in 1820. The ultra-thin crescent of the Sun's disk that forms as the Moon just starts to blot out its light, or as our lone natural satellite begins to move away from the Sun, creates a refractive phenomenon of alternating light and dark lines. Movement of Earth's atmosphere (in other words, wind) causes them to ripple.

Knowing this, and that observing shadow bands occurs infrequently, you might desire to see them; or, better yet, to photograph them. Here's how: Mount a camera with video capabilities on a tripod. Point it at the ground. Start recording 5 minutes prior to totality and stop 5 minutes after. It's as simple as that.

You can improve your chances of capturing shadow bands by aiming your camera at a light region of ground. Better yet, some observers have stretched out a white sheet or blanket on the ground and pointed the camera at it. If shadow bands appear, this technique will reveal them. Do come prepared in case the day is windy. Have some rocks or other small heavy items to weigh down the edges of the sheet. I suggest a minimum of four weights along each edge—not simply on the corners.

Chapter 19

Get to the Center Line

In all likelihood, the most important thing you'll hear from someone speaking about the eclipse or that you'll read in any book or on any website is that, on August 21, 2017, you must get to the path of totality. It's true. As I like to say in my talks, the difference between viewing a partial eclipse and *experiencing* a total one is the difference between almost dying and dying. In other words, there's no comparison.

But not only should you experience totality, I suggest that you consider taking one further step and try your best to position yourself on the eclipse's center line. Any quality map that shows the path of the 2017 total solar eclipse will have three curved lines on it; see the Google-inspired one by Xavier Jubier that's the first blog on the "Blogs" page at www.stjosepheclipse.com for an example. The two outer lines show the northern and southern limits of totality. It's within these lines that the Moon's umbra—its dark inner shadow—falls on Earth. In other words, just like your art teacher told you in third grade: Stay inside the lines.

© Springer International Publishing Switzerland 2016
M.E. Bakich, *Your Guide to the 2017 Total Solar Eclipse*, The Patrick Moore
Practical Astronomy Series, DOI 10.1007/978-3-319-27632-8_19

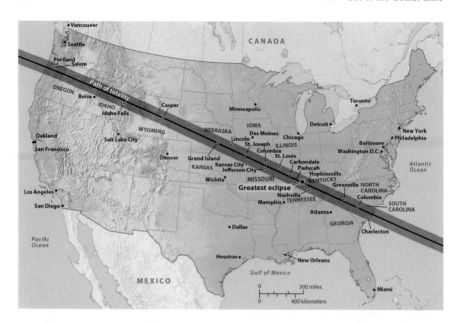

Fig. 19.1 The path of totality is the shaded region running from Oregon through South Carolina. The *black line* represents the eclipse's center line. Courtesy of *Astronomy* magazine, Richard Talcott and Roen Kelly)

But it's the line midway between those two extremes that's most important. Astronomers call this the center line, for obvious reasons. It's along this path that the central part of the Moon's shadow falls, and that's where you should try to be on eclipse day.

Here's why. Because the Moon is spherical in shape, its shadow is round. During total solar eclipses, the round shadow falls on Earth's surface. It's then your choice where you stand under the shadow.

Imagine for a moment an image of the Moon with two lines drawn through it—one passes through the Moon's center, and the other is parallel to it, but only half as long as the first line. We know the shadow cast by our nearest satellite has the same shape as the Moon itself, so you'll enjoy a longer duration of totality if the shadow traces the longer line through your location than you will if it traces the shorter one. If the duration of totality on the center line you imagined is, say, 2 minutes, you'll only experience 1 minute of totality along the other line.

Fig. 19.2 This graphic illustrates why you should try to position yourself on the eclipse's center line, represented by the *red arrow*. If you're near the northern or southern limit, represented by the *yellow arrow*, the duration of totality will be much less. (Moon image courtesy of John Chumack; graphics by Holley Y. Bakich)

If you take this example to the extreme, you could select a position on Earth that lies at the edge of the path of totality. That position would place you along either the northern or southern line I mentioned earlier. At such a location, the duration of totality would be the briefest moment, much less than 1 second. And, in fact, some observers will position themselves at the shadow's limit to record the irregularities along the Moon's limb (that is, its edge). These observations are possible only because a tiny percentage of the Sun's disk becomes visible shining through valleys or between mountains. It's important work, but it's a job for scientists. You, as a first-time eclipse viewer, want to maximize your time under the umbra.

So, get to the center line!

Now I want to say a few words about the other extreme. I've encountered several people who are bound and determined to see the eclipse from the point where totality is longest. That site, Giant City State Park, sits on 4,055 acres in Jackson and Union counties, Illinois. The park's area works out to roughly 6½ square miles. The nearest town is Makanda, whose population according to the 2010 U.S. Census, was 561. Makanda lies 7.7 miles due south of Carbondale, a city of nearly 26,000 inhabitants.

I'm concerned that vast numbers of people may head to Giant City State Park for the eclipse, and I'm not certain the facility can handle the quantity that will come. First, let me dispel the rumor that the park is 4,055 acres of pristine eclipse-viewing real estate. It isn't. It's really beautiful, that's true, but we're talking about a heavily wooded area full of hiking trails, waterways, rock formations, and some 50 types of large trees. A lot of that acreage you can't even get to, and a lot that you can access isn't great for watching the Sun-Moon dance high in the sky.

Fig. 19.3 Giant City State Park near Makanda, Illinois, is a great place from which to view the eclipse. But how many people can it hold? (Courtesy of Wikimedia Commons/Alanscottwalker)

Observers with campers wishing to get to the park early may find that all available spaces (there aren't that many) are booked. Access in and out of the park comes via Giant City Road, which orients north to south. From the mall in Carbondale (where the Wal-Mart stands just to its north on the corner), you have roughly a 12-mile drive south to the park entrance. I can't imagine what traffic might be like on this road on eclipse day, but it won't be pretty, and it won't be the optimal experience for you, your family, and friends, to have the best possible experience on eclipse day.

To help you do that, I have figured out the other locations you can travel to and still enjoy the maximum length of 2 minutes and 40 seconds of totality. To use these calculations, all you have to have is a map that shows the location of the center line. You'll find a zoomable, highly detailed one on the Front Page Science website at www.stjosepheclipse.com/map.html.

Where along the center line can you experience the maximum duration? Believe it or not, there are a lot of places that have the same time of totality—2 minutes and 40 seconds—as Giant City State Park. Totality lasts 2m39.9s (just one-tenth second less than maximum) at the intersection of the center line and Hwy 185 south of Beaufort, Missouri. At all points along the center line southeast of there, totality lasts 2m40s until you get to the intersection of U.S. Route 79 and the center line just southwest of Allensville, Kentucky, where the length of totality once again falls to 2m39.9s. That means the Moon's shadow will cover every location between Beaufort, Missouri, and Allensville, Kentucky,—a straight-line distance of 255 miles—for the maximum of 2 minutes and 40 seconds.

Now, in case you can't get to any point along that line, I went ahead and figured out the two spans along the path where totality lasts just 1 second less than maximum, and also the two parts of the center line's path where totality lasts 2 seconds less than maximum.

Two stretches along the center line have lengths of totality greater than 2m39s, but less than 2m40s. The more northwesterly stretch lies completely within the Show-Me State. It begins at the intersection of the center line and County Road 603 in Norborne, Missouri, and ends at the intersection of the center line and Hwy 185 south of Beaufort, Missouri. The southeasterly stretch begins at the intersection of U.S. Route 79 and the center line just southwest of Allensville, Kentucky, and ends where the center line intersects County Road 304 just southwest of Ten Mile, Tennessee.

Likewise, northwesterly and southeasterly stretches exist along the center line where the length of totality is greater than 2m38s, but less than 2m39s. The northwestern stretch begins just a bit east of where the center line crosses Raccoon Road in Hiawatha, Kansas, and ends at the intersection of the center line and County Road 603 in Norborne, Missouri. The southeastern stretch begins at the intersection of the center line and County Road 304 just southwest of Ten Mile, Tennessee, and ends just west of where the center line crosses Hale Ridge Road southwest of Pine Mountain, Georgia.

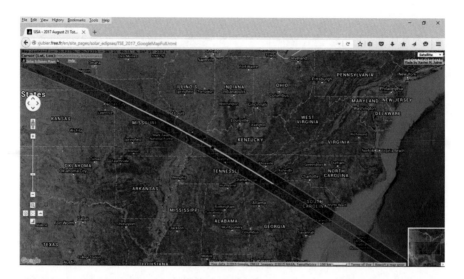

Fig. 19.4 The segment of the center line marked in *yellow* shows where you could see totality within 2 seconds of the maximum duration. (Map courtesy of Xavier M. Jubier; Data courtesy of Google/INEGI Imagery/NASA/TerraMetrics)

What all this means is that, if you position yourself along the eclipse's center line, you can experience the maximum duration of totality minus no more than a measly 2 seconds from Hiawatha, Kansas, to Pine Mountain, Georgia, a straight-line distance of nearly 766 miles! To give you some idea of how long this is, if you placed a person (with an approved pair of solar glasses, of course) every 6 feet along this stretch of center line, it would form a queue of more than 674,000 people.

Keep all this in mind. If you determine that either the cost or the effort will be too great to get to the magic spot, plenty of other locations will be out there with similar durations of totality. Indeed, there's enough shadow for us all to have a spectacular time.

Chapter 20

Start Planning for Eclipse Day

Follow the following common-sense tips, and you'll be ready.

Take Eclipse Day Off—Now!

You may think a year is a bit of a long lead time, but planning always pays, and many people near the zone of totality may have the same idea of taking off, and you don't want them to have first dibs. If not now, figure out the earliest date that makes sense for you to request August 21 as a vacation day, and mark it on your calendar.

© Springer International Publishing Switzerland 2016

229

M.E. Bakich, *Your Guide to the 2017 Total Solar Eclipse*, The Patrick Moore Practical Astronomy Series, DOI 10.1007/978-3-319-27632-8_20

Fig. 20.1 Save the date! (Courtesy of Holley Y. Bakich)

Make a Weekend Out of It

Eclipse day is a Monday. Lots of related activities in locations touched by the Moon's inner shadow will occur on Saturday and Sunday. Find out what they are, where they're being held, and which you want to attend, and make a mini-vacation out of the eclipse.

Attend an Event

You'll enjoy the eclipse more if you hook up with like-minded people. If you don't see any special goings-on a few months before August 21, call your local astronomy club, planetarium, or science center. Anyone you talk to is sure to know of eclipse activities. Travel agents also are offering trips to exotic locations that will allow you to experience the full social impact of the eclipse.

Fig. 20.2 Events like cruises to exotic locations will allow you to experience the full social impact of the eclipse. (Photo courtesy of the author)

Get Involved

If your interests include celestial events and public service, consider volunteering with a group putting on an eclipse event. You'll learn a lot and make some new friends in the process.

Watch the Weather

Meteorologists study a chaotic system. Nobody now can tell you with absolute certainty the weather a specific location will experience on eclipse day. And don't get too tied up in the predictions of cloud cover you'll see for that date. Many don't distinguish between few (one-eighth to two-eighths of the sky covered), scattered (three-eighths to four-eighths), or broken (five-eighths to seven-eighths) clouds and overcast. Also, many of the predictive websites use satellite data, which detects much more cloudiness than human observers. In both cases, you need to dig deeper.

Fig. 20.3 If on eclipse day your sky is as blue as in this photograph, you'll have no problems at all. (Courtesy of Holley Y. Bakich)

Stay Flexible on Eclipse Day

Unless you are certain August 21 will be clear, don't do anything that would be hard to undo in a short time. For example, let's say you're taking a motor home to a certain city. You connect it to power, hook up the sewage hose, extend the awnings, set up chairs, start the grill, and more. But if it's cloudy 6 hours, 3 hours, or even 1 hour before the eclipse starts, you're going to want to move to a different location. Think of the time you would have saved if you had waited to set up. Also, the earlier you make your decision to move, the better. Just imagine what the traffic might be like on eclipse day.

Stay Focused

Totality will be the shortest two and a half minutes of your life. All your attention should be on the Sun. Anything else is a waste. And be considerate of those around you. Please, no music.

Pee Before Things Get Going

Yes, this statement could be phrased more politely, but you needed to read it and follow it, too. This tip, above and beyond any other on this list, could be the most important one for you. Don't wait until 10 minutes before totality to start searching for a bathroom. Too much is happening then. Make a preemptive strike 45 minutes prior.

Record the Temperature

A point-and-shoot camera that takes movies will let you record the temperature drop. Here's a suggestion: Point your camera at a digital thermometer and a watch, both of which you previously attached to a white piece of cardboard or foamcore board. Start recording video 15 or so minutes before totality and keep shooting until 15 minutes afterward. The results may surprise you. For more about this and other projects you can do, see Chapter 18.

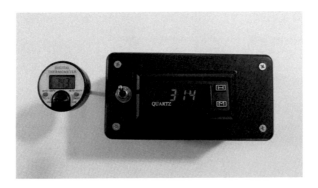

Fig. 20.4 A watch and a thermometer will help you record the temperature drop as the Moon covers the Sun's brilliant disk. (Photo courtesy of the author)

Watch for the Moon's Shadow

If your viewing location is at a high elevation, or even at the top of a good-sized hill, you may see the Moon's shadow approaching. This sighting isn't easy because as the shadow crosses St. Joseph, Missouri, for example, it is moving at 1,584 mph (2,550 km/h), or twice the speed of sound. Another way to spot the shadow is as it covers thin cirrus clouds if any are above your site. Again, you'll be surprised how fast the shadow moves.

View the 360° Sunset

During totality, take just a few seconds to tear your eyes away from the sky and scan the horizon. You'll see sunset colors all around you because, in effect, those locations are where sunset or sunrise are happening.

Get a Filter in Advance

Cardboard eclipse glasses with lenses of optical Mylar cost about $2. Such a device—it's not a toy—will let you safely look directly at the Sun. It filters out most of the light, all of the dangerous infrared (heat) and ultraviolet radiation, which tans our skin. Buy one well in advance, and you can look at the Sun anytime. Sometimes you can see a sunspot or two. That's cool because to be visible to our eyes, such a spot has to be larger than Earth. Another safe solar filter is a #14 welder's glass, which also will cost you $2. Wanna look cool at the eclipse? Buy goggles that will hold the welder's glass. I've even seen people wearing whole helmets. Either those or goggles serves one purpose—you won't need to hold the filter, so you can't drop it.

Fig. 20.5 One way to safely observe the Sun is through a #14 welder's glass (with or without the goggles). (Courtesy of Holley Y. Bakich)

No Filter? You Can Still Watch

Except during totality, we never look at the Sun. But what if you've forgotten a filter? You can still watch by making a pinhole camera. It can be as simple as two pieces of paper with a tiny hole in one of them. (Try to make the hole as round as you can, perhaps with a pin or a sharp pencil.) Line up the two pieces with the Sun so the one with the hole is closest to it. The pinhole will produce a tiny image, which you'll want to have land on the other piece of paper. Moving the two pieces farther apart will enlarge the Sun's image but will also lessen its brightness. Work out a good compromise. As mentioned earlier, a #14 welder's glass is another way to safely observe the Sun anytime.

Fig. 20.6 A pinhole projector is easy to make and 100 percent safe. (Photo courtesy of the author)

Bring a Chair

In all likelihood, you'll be at your viewing site several hours before the eclipse starts. You don't really want to stand that whole time, do you?

Don't Forget the Sunscreen

Most people who go outside during the summer know this. Remember, you'll be standing around or sitting outside for hours. You may want to bring an umbrella for some welcome shade. And if you see someone who has forgotten sunscreen, remember to share.

Take Lots of Pictures

Before and after totality, be sure to record your viewing site and the people who you shared this great event with. Speaking of sharing—a Sunspotter solar telescope by Learning Technologies (and sold through a variety of dealers) provides a sharp image of the Sun that many people can view simultaneously.

The Time Will Zoom by

In the August 1980 issue of *Astronomy* magazine, author Norm Sperling contributed a "Forum" in which he tries to convey how quickly totality seems to pass. You can read it online by heading to www.stjosepheclipse.com/Sperling8seconds.pdf.

ASTRONOMY
FORUM

Sperling's Eight-Second Law*

Everyone who sees a total solar eclipse remembers it forever. It overwhelms the senses, and the soul as well — the curdling doom of the onrushing umbra, the otherworldly pink prominences, the ethereal pearly corona. And incredibly soon, totality terminates.

Then it hits you: "It was supposed to last a few minutes — but that couldn't have been true. It only seemed to last eight seconds!"

This effect frustrated my first four eclipses, and most fellow eclipse fanatics assure me they've been bothered by it, too. Yet tape recordings, movies, and the whole edifice of astronomy and celestial mechanics all claim that it *did* last the full, advertised, two to seven minutes — to within a few seconds, that's what *really* happened.

Where did all that precious time get lost?

Eclipse Watching

True eclipse freaks recognize only two modes of life: eclipse expeditions, and preparing for them. They'll devote a year or two to perfecting equipment: telescope, camera, weird filters, and film — sandproofed, soundproofed, rainproofed (heaven forbid!), and bug resistant. No matter what their expedition sees — or does — along the way, they'll fret about totality. Will * the * clouds * part? * Will * the * equipment * work? * WILL * WE * SEE * IT?

The partial eclipse is a tantalizing, exasperating hour and a half. Then the diamond ring forms, gleams, and vanishes, and at last they have totality. They gape in awe for just a

second, then dive desperately into the sequence, many times rehearsed, of exposures, adjustments, notations so hurried they can only be unraveled from the tape recordings afterwards.

Inevitably, totality terminates too soon, often even before the planned sequence does, and they never make it to their own hard-won free-looking phase. "But I got it on film!" they proclaim, "and I can frame that and glow at it forever — even though I only saw it through the camera's finder."

The novice and the non-astrophotographer take the "hang-loose" approach. Restless in the partial phase, they get impatient and even quarrelsome around the one-hour mark. But in the last ten minutes they can feel it: totality's a-comin'. The world is darker, oranger; shadows look queerly sharp-edged. There's a nip in the air, the birds are atwitter, and shadow bands go skittering around. The ominous umbra sweeps in, the corona unfolds, the diamond glitters and is extinguished, and "OH * MY * GOD * THAT'S * THE * MOST * BEAUTIFUL * THING * I'VE * EVER * SEEN!" They stare transfixed, all their senses open, trying to take in as much as they can. Unwilling to concede that totality can't linger past third contact, they keep staring at the emerging solar sliver long after it gets painfully bright. Finally, they must be ordered to look away. Then, limp, with shit-eating grins, they applaud, or yelp, or shuffle aimlessly and ask where the next one's gonna be and how t' get there.

Both styles of eclipse-watching yield the viewer a solid eight seconds of memory.

Go ahead: replay your own. See — it's about eight seconds long! I replayed all my images of my first four totalities in about half a

*"All Total Solar Eclipses Last Eight Seconds"

24   ASTRONOMY

Fig. 20.7 "Sperling's Eight-Second Law" first appeared in the August 1980 issue of *Astronomy*. (Courtesy of *Astronomy* magazine)

Bring Snacks and Drinks

You're probably going to get hungry waiting for the eclipse to start. Unless you set up next to a convenience store, consider bringing something to eat and drink. And keep in mind that August is warm. A cooler with ice-cold drinks is a great idea.

Be Prepared To Be the Expert

If you're planning an event or even a family gathering related to the eclipse, consider this: most likely, none of the people you encounter will never have experienced darkness at noon. You will be the impromptu teacher for those attending. A telescope equipped with an approved solar filter will help Sun-watchers get the most from the eclipse, but your own knowledge is another tool that will help amplify their experience almost as much.

Invite Someone with a Solar Telescope

In the event you're thinking of hosting a private get-together (ignoring #3 above), make sure someone in attendance brings a telescope with a solar filter. While it's true that you don't need a scope to view the eclipse, having one there will generate quite a bit of buzz. And you (or the telescope's owner) can point out and describe sunspots, irregularities along the Moon's edge, and more.

Fig. 20.8 You don't necessarily need a telescope you can look through to enjoy the partial phases. This Sunspotter served admirably during the May 20, 2012 annular eclipse. (Photo courtesy of the author)

Respect Totality

The 2017 eclipse plus the events leading up to it will combine to be a fabulous social affair. Totality itself, however, is a time that you should mentally shed your surroundings and focus solely on the sublime celestial dance above you. You'll have plenty of time for conversations afterward. A get-together with family and friends after the eclipse will help you unwind a bit and hear what others experienced during the eclipse, but the actual interval should be a period for quiet observation and appreciation, without distractions.

Schedule an After-Eclipse Party or Meal

Once the eclipse winds down, you'll be on an emotional high for hours, and so will everyone else. There's no better time to get together with family and friends and chat about what you've just experienced together.

Record Your Memories

Sometime shortly after the eclipse, when the event is still fresh in your mind, take some time to write, voice-record, or make a video of your memories, thoughts, and impressions. A decade from now, such a chronicle will help you relive this fantastic event. Have friends join in, too. Stick a video camera in their faces and capture 30 seconds from each of them. You'll smile each time you watch it.

Don't Be in a Rush Afterward

Traffic gridlock will be horrendous after the event. And the sooner you try to leave after totality ends, the worse it will be. Relax. Let the part of the eclipse between third and fourth contacts play out. Many people will view this as "what we saw before totality, but in reverse," and it's a vital part of the process which can still be viewed through solar telescopes, welder's glass, or solar filters.

Chapter 21

What Do You *Really* Have to Know?

All the information in this book—as well as what you'll find in magazines, other books, and on the Internet—is great, but let's boil it down to the basics. What do you really have to know about the August 21, 2017, total solar eclipse? As I thought about this, I came up with a Top 10 list of simple facts about the event. If you're a parent, a teacher, or just someone who wants to understand the basics without all the frills, these lists are for you.

Top 10 Things to Know

1. The date the next total solar eclipse visible from the U.S. will occur is August 21, 2017. Sear this date into your memory banks.
2. A solar eclipse occurs when the Sun, the Moon, and Earth line up, in that order, and the Moon's shadow falls on Earth.
3. To see a total eclipse, you must be under the dark part of the Moon's shadow.
4. Every location in the continental U.S. will see at least a partial eclipse.
5. A partial eclipse doesn't compare to the spectacular total phase of the eclipse. In other words, *it's all about totality*.
6. The best spots to view the eclipse (outside of weather concerns) lie on the center line.
7. The total part of the eclipse is the only time you'll see the Sun's outer atmosphere. It's called the *corona*, the Latin word for "crown."
8. Except for a filter, you don't need any equipment to watch the eclipse.
9. You must use an *approved* solar filter to view the partial phases of the eclipse.
10. You *must not use* any filter to view the total phase of the eclipse.

© Springer International Publishing Switzerland 2016
M.E. Bakich, *Your Guide to the 2017 Total Solar Eclipse*, The Patrick Moore
Practical Astronomy Series, DOI 10.1007/978-3-319-27632-8_21

Top 10 Things You Might Want to Know

1. The last total solar eclipse over the continental U.S. occurred in 1979.
2. The longest span of the total part of the 2017 eclipse will last only 2 minutes and 40 seconds.
3. The only time a solar eclipse can occur is during New Moon.
4. Solar eclipses don't occur at every New Moon because the Moon's orbit is tilted to Earth's orbit around the Sun, so at most New Moons, our satellite lies either above or below the Sun.
5. First contact is the moment the eclipse begins. Second contact is the moment totality begins. Third contact is the moment totality ends. Fourth contact is the moment the eclipse ends.
6. At least two and as many as five solar eclipses occur each year.
7. The total phase of a solar eclipse can last a maximum of 7½ minutes.
8. Astronomer's call the Moon's dark inner shadow the umbra and its light outer shadow the penumbra. Under the umbra, a solar eclipse will be total. Under the penumbra it will be partial.
9. The total parts of eclipses last for different times because Earth's distances to the Sun and Moon change, so sometimes those bodies appear larger and other times smaller.
10. The next total solar eclipse in the continental U.S. will occur April 8, 2024.

Top 10 Things You Probably Didn't Want to Know

1. As the Moon's shadow passes across the U.S., its speed varies from 2,979 mph at the point of its landfall on the Oregon coast to 1,450 mph near Sparta, Tennessee. Said another way, the shadow's speed drops from a maximum of Mach 3.9 to a minimum of Mach 1.9.
2. Every clear location within the continental U.S. will see at least 48 percent (47.71 percent, to be exact) of the Sun's disk covered by the Moon.
3. This eclipse will be the 12th total (or hybrid) solar eclipse of the 21st century.
4. This total eclipse is the first to track coast to coast across the U.S. since June 8, 1918. The Moon's shadow first touched land at the Pacific coast in Washington and headed into the Atlantic Ocean after crossing Florida.
5. This will be the 4,771st solar eclipse (of any kind) since the year 1. It will also be number 9,546 out of 11,898 solar eclipses in the 5,000-year span from 2,000 B.C. to 3,000 A.D.
6. Eclipses occur because the Sun's diameter is approximately 400 times that of the Moon, but the Sun also lies 400 times as far away.
7. The eclipse August 21, 2017, will be the 21st eclipse with some totality in the continental U.S. since the United States became a country. Of those 20 previous eclipses, 19 were total and one was a hybrid with a scant 1.3 seconds of totality (maximum) visible.

8. Technically, the last eclipse seen from U.S. soil occurred July 11, 1991. Residents and visitors to Hawaii under clear skies observed it there.
9. The saros is a span of 6,585.3211 days, after which nearly identical eclipses repeat.
10. The August 21, 2017, total solar eclipse is the 22nd eclipse (out of a total of 77) in saros series 145, which began in 1639 and which will end in 3009.

Chapter 22

What to Bring
to the Eclipse

This chapter applies to all people who will be traveling to see the eclipse, but especially those who will not be part of an organized travel group. You may observe the eclipse alone, with friends or family, or at a public event like the one at Rosecrans Memorial Airport in St. Joseph, Missouri. I thought it would be good to provide a checklist of both common and unusual items that I think readers should bring to the eclipse. Such an inventory could get out of hand quickly, so I limited it to 25 entries.

And to those planning to do this alone—of the 13 total solar eclipses I've traveled to see, only the first two were lone-wolf adventures. The other eleven excursions all involved a travel agent. Of those eleven, I would rate ten of them an A and the other one a B-. That's a pretty impressive average and one that argues in favor of joining an organized tour for many reasons.

- **Sunscreen**—When someone says, "solar safety," this is what I think of. So should you. And here's something to note: If your bottle of sunscreen is more than 3 years old, replace it. That's the standard shelf life for this product.
- **Water**—August 21 will be warm everywhere in the U.S. and hot in many places. Even large events may run out of this vital fluid. Don't leave home without at least a case of bottled water.
- **Approved solar filter**—Whether you use eclipse glasses, a homemade filter using solar Mylar, or a #14 welder's glass, you will need this to view the partial phases. Also, if you plan to view the partial phases through any equipment (binoculars, telescope, etc.), you will need approved solar filters for each of them.
- **Camera**—You'll want to document the day and the activities surrounding the eclipse. My advice remains firm, though: Do not photograph the eclipse!
- **Tripod**—If you're bringing a camera, that is.

© Springer International Publishing Switzerland 2016
M.E. Bakich, *Your Guide to the 2017 Total Solar Eclipse*, The Patrick Moore
Practical Astronomy Series, DOI 10.1007/978-3-319-27632-8_22

- **Binoculars**—A great way to get a close-up view of the corona. Learn all about binoculars in Chapter 13.
- **Your eclipse guide**—OK, *this* eclipse guide.
- **Food or snacks**—Certainly this isn't as critical as water; I mean, you're not going to starve. Having some healthy snacks or some pre-made sandwiches can help you avoid fast food and give you more options in more culinary-challenged communities.
- **Medicine**—Be sure you have any prescriptions you need to take with you. And some pain medication also is a good idea. Sometimes too much Sun gives certain people headaches.
- **Toilet paper**—Let's see, millions of people on the road, rest stops few and far between … You fill in the details.
- **Hand sanitizer**—For the same reason as the last item.
- **Chairs**—Bring at least one chair (fold-up varieties pack best) for each person in your party. You're not going to want to stand for (a minimum of) 3 hours. The best chairs you can bring offer the choice of sitting upright or reclining.
- **Extra eyeglasses**—You won't forget the ones on your face, but something may happen to that pair.
- **Kid stuff**—I have no children, so I can't specify items. I can, however, advise you to bring whatever you will need to keep your offspring happy, comfortable, and occupied. You may discover, much to your chagrin, that your young children do not share your appreciation or awe for the eclipse. Don't worry, there's always 2024.
- **A broad-brimmed hat**—This will keep the Sun off your head and face, and also your neck if the hat's brim is wide enough. You'll probably sweat, but that's a great trade-off. Keep drinking water.
- **Power inverter**—You can't plug most laptops or video players directly into a car. A small DC-to-AC power inverter will let your passengers play games or movies for the whole length of the trip without having to worry about draining the batteries in their devices. Another similar device is a car-lighter-plug-to-USB-socket. Such adapters can power items that don't require much power.
- **A pillow**—Actually, bring a pillow for every chair you bring. Your passengers also can use these in the car.
- **Sunglasses**—Remember, despite their name, sunglasses are not for viewing the Sun through. They are for providing eye comfort when you look at everything else.
- **Cash**—If you meet me at the St. Joseph event, you can thank me with this. Seriously, some places may not take credit/debit cards and with the huge numbers of people in transit, it may save you some serious time just paying with cash.
- **Insect repellant**—The further down the shadow's path toward the southeastern U.S. you set up, the more important this item will become.
- **Phone**—I list this item mainly for completeness. Does anyone ever forget his or her phone anymore? Certainly nobody under 30. Now, permit me one further note about this item. It's possible that at large events (especially in smaller towns) nearby cell towers will be overwhelmed by the number of people trying

to access their cellphones. Be sure to tell anyone tracking your movements that you may be out of touch for a significant amount of time.

- **Telescope**—Be sure also to bring the minimum number of items to go with it. I won't detail them here because everyone's scope "kit" is different.
- **Astrophoto gear**—If you're going to photograph the eclipse, make sure you have the essentials you need. Keep these items together. Check them twice, then have someone else check them while you watch.
- **Odd parts and tools**—If you have a telescope, you understand what this item means. As an example, some of the things my kit contains are extra knurled knobs, an Allen wrench set, half a dozen small zip-lock plastic bags, at least two each of three types of small clamps, a micro screwdriver set (I also can work on eyeglasses with this), lens cleaning paper, at least a dozen each of two sizes of plastic zip ties, extra hardware for any tripod-mounted setups I may attempt, extra solar filters, and, you guessed it, duct tape.
- **Personal items**—You know, we won't be hiking the Himalayas or venturing into the deep ocean aboard a submarine. You'll have room for a few extras. If there's something that's especially meaningful to you and you want to bring it along, no harm done.

Chapter 23

Community Eclipse Planning

At the American Astronomical Society's August 2015 workshop concerning the total solar eclipse that will sweep the U.S. in 2017, psychologist and eclipse consultant Kate Russo announced that she had produced a white paper to help communities along the path of totality prepare for the event and that she was making it available to anyone for distribution free of charge. I found it compelling and well organized, so I reproduce it here with her permission. You also will find it in the "Blogs" section at www.stjosepheclipse.com. I ask Kate's and your indulgence if I have introduced any errors due to my minor edits.

Introduction

The purpose of this document is to provide introductory guidance for the planning of a total solar eclipse in your community. It includes an overview of the complexities of planning for this once-in-a-lifetime event in your region. The three core messages are:

1. Start early.
2. Focus on the community in addition to eclipse tourists.
3. Consult with eclipse experts to prepare for the unknowns.

When an eclipse occurs in your community, residents and visitors alike will remember it for a lifetime. Having been involved in community eclipse planning for several years now, both within my own community in Australia in 2012, and then as the lead eclipse consultant in the Faroe Islands for 2015, I know from

© Springer International Publishing Switzerland 2016
M.E. Bakich, *Your Guide to the 2017 Total Solar Eclipse*, The Patrick Moore
Practical Astronomy Series, DOI 10.1007/978-3-319-27632-8_23

personal experience that it is a challenging, exciting, and hugely rewarding role. I hope this guidance helps you with your planning.—Dr. Kate Russo, Eclipse Consultant

Fig. 23.1 No experience tops that of viewing a total solar eclipse. Here, Kate Russo celebrates after a successful sighting during the 2005 event. (Courtesy of Kate Russo)

A Note from a Past Eclipse Coordinator Tórstein Christiansen, Faroe Islands, 2015

"This document is an important one to bring to future eclipse organizers. When planning [for such an event], you have an idea of what a total solar eclipse is like. But it is not until you meet eclipse-chasers who share the actual experience that you really get an idea of what is involved in preparing your region. Kate's involvement in our preparation made us realize the importance of providing information and interaction with the community through the media. This was one of the most enjoyable aspects of planning, and the impact on the community has been so positive."

Eclipse Facts

- A total solar eclipse occurs somewhere on Earth once in every 18 months on average.

- In any one location, a total solar eclipse is rare, occurring on average once every 375 years.
- The Moon's dark inner shadow only intersects Earth along a thin track known as the path of totality.
- If you are located within the path of totality, you will experience nature's most amazing spectacle—a total eclipse of the Sun.
- For many miles on either side of the path of totality, people will experience a partial eclipse, an event nowhere near as dramatic as a total eclipse. Organizers should encourage those living outside of the path of totality with the means to travel to where the eclipse is total to do so.
- Even those who know what's happening can be caught off guard by a total solar eclipse. It is eerie, awe-inspiring, unsettling, beautiful, and often emotionally overwhelming. Most people find it hard to describe.
- It is essential to consider eye safety when planning for the eclipse. The *only* time anyone can look at the Sun safely is during the total phase of the eclipse, a duration of only a few minutes. Eclipse planners should be certain that inexpensive solar viewers will be available across the region. People must use such a filter during the partial phases of the eclipse or any other time they observe the Sun.
- Few people that you will meet have experienced a total solar eclipse. They are, therefore, unaware of how incredible this natural event is. Even those involved in eclipse planning are unlikely to have seen a total eclipse, making it a challenge for effective planning. Often the eclipse is the single largest event organized within a region in decades, attracting major crowds and international media interest on a scale never previously experienced.

Planning for the Unknown

Eclipse planning usually occurs in regions that have no living memory of seeing a total solar eclipse. Even the planners usually have never experienced the phenomenon. The community, therefore, will not know what the eclipse is, what it means for them, and how they should plan. Because of this situation, it is useful to consider a framework of the knowns and unknowns, as former U.S. Secretary of Defense Donald Rumsfeld famously referred to during a June 6, 2002, press conference at NATO headquarters in Brussels, Belgium: "There are known knowns. These are things we know that we know. There are known unknowns. That is to say, there are things that we now know we don't know. But there are also unknown unknowns. These are things we do not know we don't know."

The table below is a clearer way to frame the process of planning for things that are beyond our personal experience or awareness. It is common to simply focus planning on what is known, but effective planning is all about reducing and managing the unknowns. Those who have never experienced totality before cannot know about the eclipse experience or the needs of the community and eclipse tourists. These are the unknown unknowns. For this reason, I recommend all groups consult with eclipse experts to help them prepare for the event.

Early Planning

An eclipse is a known event (a known known) that communities become aware of years in advance. If a region decides to be a focal point of eclipse celebrations, planning needs to start well in advance. Smaller celebrations need less preparation time and those groups may wait until only a few months before the event. For larger endeavors, however, 6 months is too late to take full advantage of the opportunities to promote the region to a wide audience. In hindsight, most people who helped with planning such an event have stated that the eclipse was more significant than they originally thought.

Funding

It can be difficult to secure money for eclipse planning activities because most funding sources are unaware how significant the event will be. Regions within the path of totality benefit substantially, both in the short and long term, from the sheer scale of visitors and the huge international media interest. Local and government funding are essential to facilitate eclipse planning. Both in 2012 and 2015, I heard repeated comments about the government leaving all things up to tourism folk, and how decision were delayed without their involvement. They only realized too late how big the events were going to be and then got involved.

Funding fills several needs: (1) It ensures that the region receives wide promotion prior to this unique opportunity; (2) It helps secure venues, signage, printed materials (both for teachers and for the public), and other necessities; (3) It can fund the purchase of safe eclipse-viewing glasses, which organizers can distribute to individuals; and, perhaps most importantly, (4) It can establish a dedicated eclipse planner for the region.

Community Eclipse Planning

Organizers often view the eclipse, at least initially, as a tourist event only. However, this rare natural event that occurs within a community. It brings people together and leaves the region with a "feel good" factor. Local council and government involvement is essential to facilitate arrangements. Eclipse planning requires a dedicated person whose time requirements increase as the eclipse draws near. Additional support staff may also be necessary, especially in the final months. These include people for marketing support, managing and updating websites, media communications, and the development and production of additional materials. The eclipse coordinator typically works in a tourism capacity.

Education and Outreach

Few people have experienced a total solar eclipse, so organizers need to educate the public about what to expect. This can involve simple handouts, radio and television announcements, social media participation, and more. Teachers must have access to more robust teaching materials. In addition, organizers can schedule lectures, information events, citizen science projects, and workshops. The eclipse is an opportunity for people to come together and learn about the workings of the universe.

Fig. 23.2 Eclipse planner Kate Russo knows how to get the word out. Here, she addresses community officials prior to the November 13, 2012, total solar eclipse in Australia. (Courtesy of Kate Russo)

The Official Website

We live in an online age. Because of this fact, one of the few "must-haves" for the event is an official eclipse website and social media page. Such resources serve as the central source of information for locals and tourists. General information must appear first, then organizers can add additional material—advice on how to prepare, safe viewing techniques, road closures, and a list of related events. The website also will provide local officials a way to communicate decisions. Your official eclipse website and social media pages must be among your first key tasks. Then, update them daily.

"Negative" Media

The general public often does not see how an eclipse is relevant for them. Continual negative coverage of unrealistic numbers, traffic gridlock, food shortages, outrageous prices, estimated figures for the local economy, and scientists flocking to the region can all be off-putting for locals to constantly hear about, and potentially damaging to the tourism reputation of the region. Also, many people cannot relate if the information is solely communicated by academics or scientists. Stories about all facets of the eclipse experience that feature ordinary eclipse-chasers are important to bring the experience alive so that people can relate to the event.

Media Communication Plan

Media coverage prior to and during the eclipse is one of the greatest benefits for regions within the path of totality. Having a media communications plan that suggests educational and positive story angles can be helpful. Local, national, and international media will be looking for eclipse-related stories, and will want to interview key organizers. Along with factual information about the eclipse, those people addressing the media must have details about tourist activities in the region, numbers expected, the status of hotel bookings, the range of events available, information about the weather, details about interesting people in the region, and more.

Media Packs

The months leading up to the event provide an ideal time to generate additional material to promote your region. Prepare eclipse-related media packs, and be ready to distribute them around 3 months before the eclipse.

Media Conference

Those playing key roles in planning will find that they will be in demand with the media, especially during the 2 weeks before the eclipse. Those most in demand will be the eclipse planners, tourism officials, meteorologists, and astronomy experts. Plan a structured media conference each day starting at least 3 days before the eclipse to make this busy time easier for everyone involved.

Official Eye Safety Guidance

One of the biggest challenges for eclipse planners is managing communication about eye safety. Some individuals or groups will suggest that there is no safe way to view the eclipse. Those advising to watch the eclipse on TV, to remain indoors, or to turn their backs to the event are uninformed, and may simply with to avoid possible litigation, rather than to educate about safe viewing.

Fig. 23.3 Eclipse chaser Kate Russo stresses the importance of eye safety at every opportunity. She is never without her approved eclipse glasses. (Courtesy of Paul McErlane)

Inconsistent information in the past has led to confusion and the unfortunate result of people not viewing the magnificent event. Eye safety advice is available from reputable online sources, and many inexpensive methods of viewing the Sun exist. Plans to communicate them must be in place well in advance of the event to ensure consistency and avoid confusion.

Strategy

Planning for the eclipse initially tends to focus on potential tourists because those in the tourism industry lead eclipse planning usually. However, in the U.S. in 2017, in all but the smallest communities, the local population will make up the largest numbers of those viewing an eclipse. Locals and tourists differ significantly in how they approach the eclipse, and both groups must be taken into account. The community will remember this event, so plans must ensure community involvement.

The local population most likely will want to view this special event in places that are meaningful for them. Some choose to do this in a large public gathering. Others will choose to do this at home with their family and friends. Eclipse chasers, on the other hand, are not committed to view from any one location and will favor mobility, confirming their viewing plans the day before or even the day of the event based on the weather forecast. Planners can reduce the number of people necessary to run events by using locations or facilities with large capacities. Such venues also foster the idea of community.

The largest group of visitors will be those living within a half days' drive of the path of totality, who will travel to the community just for the eclipse. Such people will be looking for public viewing locations. Visitors from afar usually stay in the region of an eclipse for about 3–5 days, or longer if they have traveled a great distance. So, for the Indonesian eclipse March 9, 2016, eclipse-chasers generally will spend a week or more in Bali, the top tourist destination in the region.

In 2017 in the U.S., 3 days will be more likely because this eclipse occurs on a Monday. A late arrival Friday, therefore, will make sense to many people. Visitors will want to view the eclipse, of course, but also participate in a range of celebratory and tourist activities. If merchants in your community artificially raise prices too high, people will simply choose to stay elsewhere, choosing to drive in on the day.

Information about any planned road closures or parking restrictions must be communicated well in advance. It is essential to make plans for traffic management and parking to facilitate the movement of large numbers on eclipse day.

Weather Updates

The path of totality usually covers a narrow but long path, resulting in people having a choice of locations for viewing. The best places for viewing are anywhere along the center line and in locations that have the best chance of clear skies at eclipse time. All eclipse chasers will gladly sacrifice seconds of totality for a better chance of clear skies. Regions with superior climate at eclipse time will have a distinct advantage and should already be making plans to appeal to eclipse tourists.

The weather on eclipse day is of utmost importance. The eclipse will happen regardless, but if it is cloudy then nobody will see it. A cloudy eclipse does not provide any of the awe that seeing totality does, although it still will provide a memorable moment. Organizers should be aware that changes in weather will mean people will uproot themselves to travel to—or from—your location. Most U.S. meteorologists already have some astronomical knowledge, so they will convey eclipse-related weather information in a practical and timely manner so people can make informed choices regarding their viewing locations.

Leadership

Ideally, a community will appoint an Eclipse Coordinator as the go-to person for everything related to the eclipse. This person needs to have project management experience and ideally will have existing relationships with a wide variety of tourism and government stakeholders, who they will lead throughout the process. They need to learn the facts about the upcoming eclipse, and to be confident at communicating them effectively at meetings, events, and to the media. Eclipse Coordinators need to view the event as significant for the community and to convey this belief to others.

The Eclipse Coordinator must be proactive and strategic regarding the many decisions and actions that need to be taken. Often, they must suggest actions on things that are outside of their usual control. Examples are prompting consideration for eclipse day being a public holiday; whether schools should close to facilitate families wanting to share the event; arranging for automatic outdoor lights to be turned off during the eclipse; and encouraging agreement across the region regarding capping prices to avoid a negative reputation and visitors choosing to view elsewhere. They may have to encourage others, such as local businesses and artists, to find ways to benefit from the eclipse opportunity through creating local eclipse-related merchandise.

Eclipse Task Force

I strongly urge your community to develop an Eclipse Task Force, which must consist of a range of stakeholders from across the region. These include, but are not limited to, representatives from tourism, government, policing, event coordination, creative industries, education, health, business, and local media.

I strongly recommend that a community supplements its Eclipse Task Force with expert advisors in order to reduce the unknowns. The following are key advisors to consider with their potential roles. In some regions, credible individuals can fulfill several of these roles.

Astronomy Expert

This person will ensure that all astronomical information about the event is correct. In addition to serving on the Eclipse Task Force, this person probably will lead the community viewing event. They can advise on viewing locations, equipment, eye safety, and a variety of other concerns.

Science Educator

Along with the astronomy expert, a science educator can help develop the program of events, exhibitions, workshops, and lectures for schools and for the public. They may also develop viewing activities or citizen science projects. Usually, this person works at a local museum, science center, or school.

Weather Expert

The Eclipse Task Force must contain at least one local meteorologist. This person can provide information about past weather patterns, areas to avoid due to micro-climates, and specific eclipse weather changes. In the final days before the event, this person plays a crucial role and will be one of the most in-demand experts in the media.

Eclipse Chaser

This task force member can help address the unknown unknowns. They have experienced totality before, perhaps many times, and have first-hand experience with how other regions have prepared for a total solar eclipse. They will contribute practical advice based on their experience.

Conclusion

A total solar eclipse often imparts a long-term legacy to a region, including a significant economic benefit, new strategic partnerships, international exposure, new tourism connections, and a feel-good factor that lasts a lifetime. The event often inspires children and adults alike to develop an interest in science. Helping to plan for a total eclipse in your region is challenging, but it is also exciting and rewarding. Indeed, the awe-inspiring phenomenon that is a total solar eclipse provides a unique opportunity to promote your region to the whole world.

I hope that this narrative has highlighted both the challenges and the opportunities involved in community eclipse planning and has given preliminary guidance on how to address some of these issues. Follow this plan and it will enhance how people experience the eclipse in your region. Indeed, it will be an unforgettable experience that leaves a positive impact for years to come.

About the Author

I am an author, psychologist, and eclipse chaser, and I have served as an Eclipse Consultant for a variety of agencies. I saw my first total solar eclipse in 1999 and have now seen nine of them. As a psychologist, I am fascinated by the experience of totality and have been researching this for several years.

Fig. 23.4 Psychologist and author Kate Russo signs copies of her book, *Total Addiction: The Life of an Eclipse Chaser* (Springer, 2012) after speaking about her passion. (Courtesy of Paul McErlane)

I became interested in how communities plan for eclipses in 2012, when the path of totality occurred in my home region of North Queensland, Australia. For the first time, I was a local within the community in the lead up to an eclipse. This gave me unique insights into the local perspective—and highlighted that key eclipse messages were not getting through. I spoke to many people who did not see that the eclipse was relevant to them, with some stating they were planning to leave the region to "avoid the chaos."

I then went to work doing as much outreach as I could to ensure that my fellow locals knew the eclipse wasn't just for tourists or scientists, but rather a special event for the whole community. I was already interviewing locals before and after the eclipse, and I included eclipse planners in these interviews to capture the planning process and lessons from hindsight.

I then went on to become the Eclipse Consultant for the total solar eclipse March 20, 2015, in the Faroe Islands, ensuring that both the local community as well as tourists were prepared for the event. After the eclipse, I again interviewed those involved in planning to gain further insights into the planning process. This report is the result of what I have learned through all of these activities over those years. I am keen to share my experiences with others who are lucky enough to be living within a future path of totality and to help them prepare their communities for a wonderful event.

— Kate Russo, Eclipse Consultant
email: kate@beingintheshadow.com
Facebook: www.facebook.com/BeingInTheShadow
Web: www.beingintheshadow.com

Chapter 24

Detailed Weather Predictions Along the Center Line

The initial summary in this chapter comes from the website of highly regarded Canadian eclipse meteorologist Jay Anderson (www.eclipser.ca). He states: "Material on this web page, except where identified as belonging to others, may be copied freely and used with or without attribution. I appreciate acknowledgement, but it is not a necessary condition for use of these graphics and tables. Higher-resolution copies of many of the maps and diagrams are available on request." I reproduce Anderson's text here with only the mildest of edits.

I have always had the utmost regard for Jay. Through the last several decades, I have communicated with him a number of times asking for his advice or help answering a question. I finally got to meet him at an eclipse conference in Portland during August 2015. There, I peppered him with a volley of questions that would make a Hollywood actor blanch. But he answered each of them accurately and with good humor.

Anderson first produced a study of weather prospects for the solar eclipse February 26, 1979, after he realized he was living in the path of totality. He repeated the research for the total solar eclipse that took place November 22, 1984, and he's been doing it ever since. Anderson retired as a meteorologist a few years ago, but still teaches courses in meteorology, climate change, and storm chasing at the University of Manitoba. He also served as editor of the *Journal of the Royal Astronomical Society of Canada* from 2006 to 2015.

© Springer International Publishing Switzerland 2016
M.E. Bakich, *Your Guide to the 2017 Total Solar Eclipse*, The Patrick Moore
Practical Astronomy Series, DOI 10.1007/978-3-319-27632-8_24

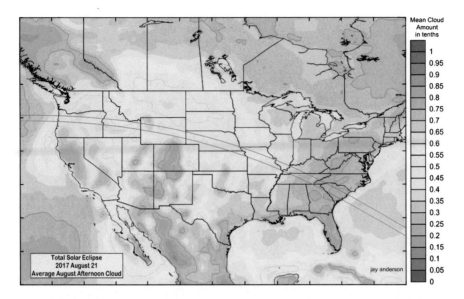

Fig. 24.1 This eclipse track map shows the average afternoon August cloudiness that meteorologist Jay Anderson derived from 22 years of satellite observations. He outlined the 50 and
75 percent contours. The best prospects for seeing the eclipse are in Oregon and Idaho. Chances
decrease steadily as you move toward the southeast. (Map courtesy of Jay Anderson; data
courtesy of CIMSS/NOAA/UW-Madison)

Early Weather Prospects for the 2017 Eclipse

This eclipse arrives at a propitious time: the summer thunderstorm season is wind-
ing down and retreating southward; the Arizona monsoon is breaking; and the
storm-carrying jet stream has not yet begun its journey southward from Canada.
The dry and generally sunny fall season is about to begin. After a 38-year eclipse
drought, this one arrives to the open arms of a friendly August climatology. The
best weather prospects are found in the West and Midwest, but weather forecasting
has now reached a level of accuracy that movement to a favorable area can be
planned several days or a week in advance.

The westerly winds that bring weather systems onto the North American conti-
nent first have to cross the mountain barriers that border the Pacific coast. The
moisture-laden air is forced to rise up the windward slopes, giving seaward-facing
Washington and Oregon unfortunate reputations for cloud and precipitation. It's an
undeserved bad rep, as the cloud is confined mostly to the coast and only a short
move inland will bring some of the best weather of the track. To top it off, August
is not really a wet and cloudy month to begin with. The diminutive Coast Range is
sufficient to erase the cloud from the sky, giving the Willamette Valley and the city
of Salem some fine eclipse weather.

East of Salem, the eclipse track crosses the Cascade Mountains and the cloudiness bumps up a touch—not enough to detract from the appeal of watching an eclipse from the majesty of a beautiful range of mountains. Once over the Cascades, the eclipse path moves onto the Columbia Basin, where it finds the best weather conditions anywhere along the track.

Even though the summer months are already notable for their abundant sunshine, the Cascade mountains manage to extract the little moisture remaining in the Pacific westerlies, giving Madras, Oregon, and its surroundings along the Deschutes River a very low average cloud cover, as seen in both the satellite cloud observations and the surface-based observations at Redmond in the table. In the recent past, few eclipse tracks have been able to offer such a low average cloudiness.

East of Madras, the cloud cover rises across the central and eastern portions of the Columbia Plateau. Part of this is because of the terrain, which rises into a rougher series of small mountains, but the area is also the breeding ground for summer thunderstorms when enough moisture creeps into the plateau. Over the eastern Plateau, midsummer precipitation is about 70 percent more than on the leeward slopes of the Cascades near Madras, but both are noted for their dry desert-like climate.

The Columbia Plateau is interrupted along the Idaho state line by the Snake River Valley. Once again, the westerly Pacific winds must drop to lower altitude, drying adiabatically as they descend and creating another region of very promising cloud cover climatology near Huntington, Oregon, and later in the Payette River Valley, at Cascade, Idaho. Given the uncertainties in the measurement of cloud cover by both satellites and humans, it is impossible to say that the Oregon-Idaho boundary has less cloud than the Deschutes River Valley, as the difference is no more than a few percentage points.

Past the Snake River Valley, cloudiness follows the terrain as the Moon's path rattles across the Rockies. Each ascent is accompanied by an increase in cloudiness; each descent brings a Chinook-like drying and an increase in sunshine and eclipse prospects. The general trend in cloudiness is upward until the track finally descends onto the Great Plains for good at Casper, Wyoming.

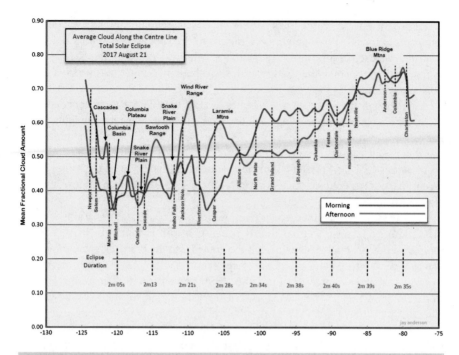

Fig. 24.2 This chart shows the average morning and afternoon cloud cover along the eclipse center line. Meteorologist Jay Anderson extracted it from 20 years of satellite imagery. Dashed vertical lines above the names of cities and towns along the track show their locations. The names of prominent topographic features lie above the graphs. (Courtesy of Jay Anderson)

Thunderstorms are the main cloud-making culprit here, as the upper slopes of the Rockies, particularly Idaho and western Montana, are a storm nursery. They foster storms that may later head out onto the plains if moisture and wind conditions are right. They also start forest fires and eclipse chasers will have to monitor and move to avoid any smoke that may make its way onto the eclipse track, possibly from as far south as Arizona and New Mexico if upper winds are out of the south. Smoke tends to have a northeastward trajectory under the usual summer conditions, so points in Oregon and western Idaho are less likely to be obscured. If smoke is present in any amount, the eclipse may adopt an unbecoming pink color, even though the Sun will be relatively high.

Though it lies on the east side of the Rockies and should have a dry climate, Casper is shown to have a relatively high cloud amount in the satellite statistics, but a very much smaller one—comparable to Madras—in the surface climatology. In this case, the surface record is preferred.

Casper lies north of the Laramie Mountains, a spur of the Rockies that runs southeast-northwest rather than north-south, the general trend of the mountain chains. Satellite measurements have a footprint of 1° (about 100 kilometers) on a side and the value at the location of Casper is capturing much of the cloud on the spur. The Laramie Mountains, covered with dark forests, have a tendency to form thunderstorms because of their low albedo, which, by absorbing sunlight, warms the slopes more strongly than the non-forested lowlands.

The ups and downs of mountain cloudiness fades away as the eclipse track moves out onto the Great Plains east of Casper and is replaced by a steady rise in cloudiness toward the Mississippi River. This trend is a reflection of the northward flow of moisture from the Gulf of Mexico that feeds the Plains thunderstorms in the warm months. From Nebraska to Missouri, average cloud cover as measured by satellite increases from about 50 percent to 65 percent.

In the surface record, the increase is from around 25 percent to 45 percent. This dichotomy likely results from the sensitivity of the satellite sensors to thin high-level cloud, but human observers are also relatively insensitive to the same sky condition, tending to observe less cloud. In both measures, the lowest cloud amount along this part of the center line is found to the west of North Platte, Nebraska.

2017 Eclipse Track August Weather Stats	area communities	POPS	Frequency of Sky Condition (%)							
			Clear	Few	Scattered	Broken	Overcast	Thin fog	Fog	Average
Oregon										
Southwest Oregon Regional	North Bend		19.8	14.2	4	13.4	48.4	0.2	0	63
Umpqua River CGS	Winchester Bay		0	30.5	20.9	48.7	0	0	0	52
Newport Municipal *	Newport		18.4	15.6	0	18.4	32.8	14.4	0.3	64
Astoria	Astoria		7.7	9.6	5	17.1	60.4	0.2	0	78
Corvallis Muni *	Corvallis		27.6	15.5	0.9	21.6	34.5	0	0	54
Mahlon Sweet Field	Eugene		36.9	13.7	5.8	14.1	29.5	0	0	45
McNary Field *	Salem		36.7	13.9	4.4	13	32	0	0	46
Aurora State *	Aurora		38.1	11.7	2.1	16.7	31.4	0	0	47
Troutdale	Troutdale		21	16.8	1.7	17.9	42.5	0.2	0	60
Portland International	Portland		22.4	12.5	7.8	18.2	39.1	0	0	59
Klamath Falls Airport	Klamath Falls		42.2	36.1	0.7	16.2	4.9	0	0	24
Roberts Field *	Redmond		50.4	18.1	7.1	15.9	8.3	0	0.3	27
Burns Municipal Airport	Burns		57.9	14.9	4.1	16.9	6.2	0	0	24
Eastern Oregon Regional	Pendleton		47.3	17	8	17.3	10.4	0	0	30
Baker Muni *	Baker City		38.3	29.8	2.3	16.8	12.9	0	0	32
Ontario Muni *	Ontario		77.4	6.5	0	4.3	11.8	0	0	16
Idaho										
Boise Municipal	Boise	85	50.3	18.9	6.1	15.2	9.6	0	0	27
McCall Municipal	McCall		35.4	24.2	0	32.3	7.1	1	0	37
Friedman Memorial	Hailey		39.8	35.3	0.3	17.1	6.7	0.3	0.5	27
Lemhi County	Salmon		34.7	35.3	0	20.5	9.1	0.3	0	31
Pocatello Municipal	Pocatello	81	43.7	21	7.5	18.1	9.7	0	0	30
Idaho Falls Regional *	Idaho Falls, Amon, Shelley		42.5	24.6	1.1	17.7	12.8	0.2	1.1	32
Allen H. Tigert	Soda Springs		38.4	39.3	0	17.9	3.6	0	0.9	25
Wyoming										
Jackson Hole Airport *	Jackson Hole/Teton Village		25.7	42.5	0	21.6	9.9	0	0.3	34
Hunt Field Airport *	Lander/Boulder Flats	75	21.6	32.3	10.3	22.5	13.4	0	0	41
Riverton Regional *	Riverton		29.5	40.5	0	21.6	8.4	0	0	32
Worland Municipal Airport	Worland/ McNutt		35.1	37	0	18.2	9.2	0	0.5	30
Natrona County *	Casper		25.5	26.7	9.5	20.3	17.9	0	0	42
Gillette-Campbell County Arpt	Gillette		29.5	19.3	17.7	24.3	8.9	0	0.2	39
Nebraska										
Western Nebraska Regional *	Scottsbluff/Minatare		25.8	22.5	11.6	24.6	15.6	0	0	43
Alliance Muni *	Alliance		32.6	32.8	0	22.5	12	0	0	35
Hooker County Airport *	Mullen		21.9	32.1	0.0	20.1	25.6	0.2	0.0	47
North Platte Regional Arpt *	North Platte	75								
Broken Bow Muni *	Broken Bow		20.1	32.4	0	27.7	19.3	0.5	0	47
Kearney Regional Airport *	Kearney/Gibbon		28	27.7	0	12.6	31.5	0.2	0	46
Hastings *	Hastings		20.6	31.9	0	32.3	15.2	0	0	45
Central Nebraska Regional Arpt *	Grand Island		25.5	14.6	14	23.4	22.5	0	0	49
Lincoln Municipal *	Lincoln	70	21.9	18.5	7.7	24.2	27.5	0.2	0	53
Beatrice Municipal *	Beatrice		16.4	31.1	0	33	19.4	0	0	50
Columbus Muni	Columbus		17.4	33.7	0	32	16.1	0.6	0.2	47
Brenner Field Airport *	Falls City		20.9	27.9	0	36.4	14.8	0	0	48
Sidney Municipal/ Lloyd W. Carr	Sidney		20.3	20	22.2	22.6	14	0.9	0	45
Chadron Municipal	Chadron		19.5	41.5	0	25.7	13.2	0	0	40
Kansas										
Forbes Field	Topeka		18.5	32.7	0	29.6	19.2	0	0	47
Johnson County Executive Airport	Olathe		15.7	34.7	0	31.4	17.1	0.8	0.2	48
Missouri										
Rosecrans Memorial Arpt *	St Joseph		20.3	23.4	2.7	29.8	23.8	0	0	52
Kansas City International *	Kansas City/Platte	67	21.8	19.3	12.1	25.9	20.9	0	0	49
Columbia Regional *	Columbia/Ashland	64	13.2	24.1	11.2	24	27.4	0	0	55
Jefferson City Mem *	Jefferson City		14.6	33.4	0	30.7	21.3	0	0	51
St Louis International Apt	St Louis		11	21.4	13.9	30.8	22.9	0	0	56
Spirit of St Louis Airport *	St Louis	65	14.8	32.3	3.9	28.6	20.3	0	0	49
Cape Girardeau Muni *	Cape Girardeau/Jackson		11	34.6	4.6	30.9	18.6	0	0.3	50
Waynesville Regional	Waynesville / Fort Leonard Wood		3.5	46.6	0	33.1	16.8	0	0	50
Illinois										
St Louis Downtown Apt/Cahokia *	St Louis		8.4	31.5	0	41	19.1	0	0	56
Abraham Lincoln Capital Arpt.	Springfield	70	11.9	17.4	15.6	29	25.9	0.2	0	58
Scott AFB/MidAmerica	Scott AFB/Shilo		7.9	44.2	0	31.4	16.6	0	0	48
Southern Illinois *	De Soto/Murphysboro		11.8	38.4	0	33.8	15.5	0	0.5	48
Williamson County Regional *	Marion/Herrin/Carterville		5.7	35.5	0	41	17.8	0	0	55
Cairo	Cairo	75								
Kentucky										
Barkley Regional Airport *	Paducah	71	8.2	20.9	16.8	34.7	19.4	0	0	57
Owensboro-Daviess County	Owensboro		11.3	37.9	0	31.6	19.2	0	0	50
Bowling Green/Warren County *	Bowling Green		6.1	33.7	0	40.4	19.6	0	0.2	57
Tennessee										
Nashville International Arpt *	Nashville	63	4.9	19.1	23.8	32.9	19.2	0	0	58
Smyrna Arpt	Smyrna		1.1	46.5	0	40.1	12.3	0	0	51
Chattanooga Metropolitan Arpt	Chattanooga		7.9	23.7	37.2	24.8	6.4	0.0	0.0	46
North Carolina										
Asheville Municipal	Asheville	54	3.6	23.4	15.2	32.7	25.2	0	0	61
Georgia										
Athens-Ben Epps Airport	Athens		3.4	21.5	20.2	36.5	18.3	0	0	59
South Carolina										
Anderson Regional Arpt	Anderson		4.9	44.5	0	34.4	15.8	0	0.3	50
Greenville Spartanburg	Greenville	61	4	20.3	17.6	37.2	20.8	0	0	61
Columbia Metropolitan	Columbia	66	2.8	20.9	19.1	34.7	22.5	0	0	61
Charleston AFB/International	Charleston	64	2.8	13.8	15.4	39.2	29	0	0	69
Beaufort MCAS	Beaufort		0.6	14.7	14.1	53.4	17.1	0	0	67

POPS=percent of possible sunshine; average cloud = arithmetical average of observed cloud amounts; * = within the umbral shadow

Fig. 24.3 This table lists weather statistics collected from surface stations located along and close to the eclipse track. Note that all weather statistics, including those from satellites, have biases and errors. Use them for the comparison of one site with another, not the absolute probability of seeing the eclipse. (Courtesy of NCDC)

As the shadow track crosses through Missouri, it moves into rougher terrain of the Ozark Plateau south of St. Louis and cloudiness climbs more steeply, reaching a maximum at Festus, just before crossing the Mississippi River. The Ozarks are not particularly high, and so the cloud cover maximum is probably related to both the albedo of the forests and the air flow that must rise over the terrain. Cloudiness falls briefly over the flat landscape of southern Illinois, but begins a steady climb to a maximum on moving into Tennessee. After passing Nashville, the lunar shadow will begin the slow climb to the crest of the Appalachians, which is reached along the Tennessee-North Carolina border. Average cloudiness rises through the whole of this part of the track, peaking as high as 80 percent by satellite or just short of 60 percent in the surface data.

The two datasets are in poor agreement in this part of the lunar path; the greatest cloud amount recorded by observers on the ground comes at Nashville, Tennessee, while the satellite record suggests a peak just short of the Atlantic Ocean, near Columbia. Of all of the states along the track, Tennessee may have the most complicated weather, as the terrain climbs from the Gulf Coastal Plain in western parts, to the Highland Rim in middle Tennessee and to the Cumberland Plateau with the Blue Ridge Mountains and all of its tributaries in the east. To complicate the terrain, the Nashville Basin provides a low-level haven in the Highland Rim, while the Tennessee River Valley does the same in the Blue Ridge Mountains.

The Appalachians are well known for their cloudiness at all seasons: on the western slopes, Gulf moisture is forced to rise; on the eastern slopes, Atlantic moisture is attracted inland and upslope. There is no substantial lee slope drying on either side of the eastern mountains, but individual valleys to benefit from the descent of air whatever the source, and so there is considerable up-and-down in the cloud averages, particularly in the surface-based observations.

The final descent to the Atlantic Ocean brings a small 7- or 8-percent improvement before the track heads out over the water. Some of this will be due to the suppression of convection by the cool breezes from the ocean.

Atlantic eclipse observers will have to be conscious of the hurricane season, which is well underway in late August. Fortunately, they are relatively uncommon on any single day, and so the likelihood of storm intervention is low. In their favour—skies tend to clear in the wake of a hurricane. Ship-board observers are likely to be more affected than those on land, as movement away from an approaching hurricane on the water will take place days in advance, and return to the track may be equally slow.

Some Thoughts on Satellite- and Human-Based Weather Observations

Satellites don't observe clouds, they observe visible and infrared radiances and at night, only infrared radiances. Depending on the algorithm used to extract cloud information, they may be biased to over- or under-report high cloud or low cloud, or perhaps both. There are at least a half-dozen global cloud datasets, and they all give different estimates for cloudiness, sometimes by a very large amount. The dataset used to construct Graph 1 is the one that most resembles the surface observations.

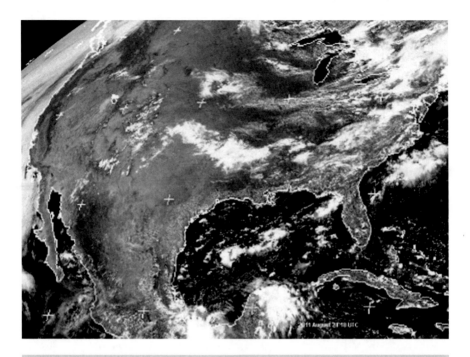

Fig. 24.4 This image shows the GOES East satellite view August 21, 2011, at 18h00m Universal Time (2 P.M. EDT). (Courtesy of NOAA Satellite and Information Service)

Fig. 24.5 This image shows the GOES East satellite view August 21, 2012, at 18h00m Universal Time (2 P.M. EDT). (Courtesy of NOAA Satellite and Information Service)

Fig. 24.6 This image shows the GOES East satellite view August 21, 2013, at 18h00m Universal Time (2 P.M. EDT). (Courtesy of NOAA Satellite and Information Service)

Fig. 24.7 This image shows the GOES East satellite view August 21, 2014, at 18h00m Universal Time (2 P.M. EDT). (Courtesy of NOAA Satellite and Information Service)

Beginning in the 1990s, human observations of cloud cover were replaced by those acquired by automatic weather stations. Those automatic observations were limited by the capability of the laser measuring systems to clouds below about 15,000 feet, so that higher clouds were simply recorded as "clear." Table 1 shows cloud cover statistics derived from 1970–1998 in order to avoid the biases introduced by the automatic measurements.

Use these data for comparative purposes, rather than absolute probabilities of seeing the eclipse. The percent of possible sunshine statistics are more reliable, but pertain to the whole day rather than the hours near maximum eclipse and are only available for a few sites. In general, human observations are 20–30 percent lower than satellite-based cloud estimates.

Summary

The United States has a sophisticated private forecast industry and a great resource in the National Weather Service. Reliable weather forecasts will be available for a week or more before the eclipse, and with a little mobility, no one who wants to travel to see this event should be disappointed.

Now let's take a look in a bit more detail about local conditions, including cloudiness, at some of the key locations along the eclipse path.

Salem, Oregon

Climate and Topography

Salem has a Mediterranean climate with dry warm summers. The area within 25 miles features *forests* (54 percent) and *croplands* (43 percent).

Temperatures

August has daily highs decreasing from 83 °F to 79 °F through the month, exceeding 94 °F or dropping below 69° F only 1 day in 10.

Cloud Cover

The median cloud cover ranges from 12 percent (mostly clear) to 18 percent (mostly clear). In mid-August, the sky is *clear or mostly clear* 62 percent of the time, *partly cloudy* 12 percent of the time, and *mostly cloudy or overcast* 25 percent of the time.

Rain

Rain is most likely around August 31, occurring in 21 percent of days. It's least likely around August 7, occurring in 16 percent of days. *Light and moderate rain* occur during 59 and 21 percent of days with precipitation, respectively. They are most likely at month's end. *Thunderstorms* occur during 12 percent of those days with precipitation. They are most likely around the beginning of the month, but occur only during 2 percent of all days.

Humidity

The relative humidity typically ranges from 35 percent to 91 percent, rarely dropping below 22 percent and reaching as high as 100 percent. The air is *driest* around August 2 (below 41 percent 3 days out of 4) and *most humid* around August 27 (above 87 percent 3 days out of 4).

Wind

Over the course of August typical wind speeds vary from 0 mph to 13 mph (calm to moderate breeze), rarely exceeding 16 mph. The *highest* average wind speed of 6 mph occurs around August 1 and the *lowest* of 6 mph around August 26. The wind is most often out of the south (26 percent of the time) and north (15 percent).

Redmond, Oregon (Closest Weather Station to Madras)

Climate and Topography

Redmond has a cold semi-arid steppe climate. The area around it is covered by *shrub lands* (52 percent), *grasslands* (26 percent), *forests* (19 percent), and *croplands* (3 percent).

Temperatures

August's daily high temperatures decrease from 86 °F to 79 °F over the course of the month, exceeding 95 °F or dropping below 68 °F only 1 day in 10.

Cloud Cover

The median cloud cover is 2 percent (clear). In mid-August, the sky is *clear or mostly clear* 67 percent of the time, *partly cloudy* 8 percent of the time, and *mostly cloudy or overcast* 13 percent of the time.

Rain

The average probability for rain on a given day is 15 percent, with little variation through the month. The most common forms of precipitation are light rain and thunderstorms. Each occurs during half of those days with precipitation. Light rain is most likely at month's end. *Thunderstorms* are most likely around the beginning of the month.

Humidity

In August, relative humidity typically ranges from 20 percent to 78 percent, rarely dropping below 12 percent or reaching as high as 98 percent. The air is *driest* around August 1 (below 23 percent 3 days out of 4) and *most humid* around August 31 (above 67 percent 3 days out of 4).

Wind

Over the course of August typical wind speeds vary from 0 mph to 16 mph, rarely exceeding 20 mph. The *highest* average wind speed occurs around August 1 and the *lowest* around August 28. The wind is most often out of the south (18 percent of the time), northwest (18 percent), and north (11 percent).

Lime, Oregon (Using Data from the Ontario, Oregon Weather Station)

Climate and Topography

Ontario, Oregon has a cold semi-arid steppe climate. The area around this station is covered by *shrub lands* (45 percent), *grasslands* (45 percent), *croplands* (6 percent), and *forests* (2 percent).

Temperatures

August sees daily highs decrease from 93 °F to 85 °F during the month, exceeding 99 °F or dropping below 76 °F only 1 day in 10.

Cloud Cover

The median cloud cover is an astounding 0 percent (clear). In mid-August, the sky is *clear or mostly clear* 93 percent of the time, *partly cloudy* 1 percent of the time, and *mostly cloudy or overcast* 4 percent of the time.

Rain

The average probability that some form of precipitation will be observed in a given day is 11 percent, with little variation over the course of the month. Throughout August, the most common forms of precipitation are light rain and thunderstorms which are observed during 59 and 35 percent of days with precipitation respectively. Light rain is most likely around August 26 (8 percent). *Thunderstorms* are most likely at the start of the month (5 percent).

Humidity

The relative humidity in August typically ranges from 19 percent to 71 percent, rarely dropping below 12 percent or exceeding 88 percent. The air is *driest* around August 1 (below 22 percent 3 days out of 4) and *most humid* around August 31 (above 64 percent 3 days out of 4).

Wind

In August, typical wind speeds vary from 0 mph to 14 mph, rarely exceeding 23 mph. The *highest* average wind speed occurs around August 1 and the *lowest* occurs around August 19. The wind is most often out of the west (18 percent of the time) and northwest (13 percent).

Terreton, Idaho (Data from the Idaho Falls Weather Station)

Climate and Topography

Idaho Falls has a cold semi-arid steppe climate. The area near this station is covered by *croplands* (38 percent), *forests* (22 percent), *shrub lands* (21 percent), and *grasslands* (19 percent).

Temperatures

August sees daily highs fall from 86 °F to 80 °F over the course of the month, exceeding 93 °F or dropping below 70 °F only 1 day in 10.

Cloud Cover

The median cloud cover is 23 percent (mostly clear). In mid-August, the sky is *clear or mostly clear* 52 percent of the time, *partly cloudy* 17 percent of the time, and *mostly cloudy or overcast* 27 percent of the time.

Rain

The average probability for rain on a given day is 22 percent, with little variation through the month. The most common forms of precipitation are light rain and thunderstorms. Each occurs during half of those days with precipitation. Light rain is most likely around August 23. *Thunderstorms* are most likely around the beginning of the month.

Humidity

In August, the relative humidity typically ranges from 22 percent to 92 percent, rarely dropping below 13 percent or reaching as high as 100 percent. The air is *driest* around August 4 (below 25 percent 3 days out of 4) and *most humid* around August 1 (above 85 percent 3 days out of 4).

Wind

Typical wind speeds vary from 0 mph to 17 mph, rarely exceeding 26 mph. The *highest* average wind speed occurs around August 31 and the *lowest* around August 1. The wind is most often out of the southwest (25 percent of the time), north (23 percent), and south (19 percent).

Jackson Hole, Wyoming

Climate and Topography

Jackson Hole has a humid subarctic continental climate with cool summers and no dry season. The area surrounding this station is covered by *forests* (85 percent), *croplands* (9 percent), *grasslands* (4 percent), and *lakes and rivers* (2 percent).

Temperatures

August's daily highs decrease from 78 °F to 70 °F over the course of the month, exceeding 83 °F or dropping below 64 °F only 1 day in 10.

Cloud Cover

The median cloud cover ranges from 15 percent (mostly clear) at the start of August to 23 percent (mostly clear) at the end. In mid-August, the sky is *clear or mostly clear* 70 percent of the time, *partly cloudy* 11 percent of the time, and *mostly cloudy or overcast* 18 percent of the time.

Rain

The probability of rain varies throughout the month. It's most likely around August 28, occurring in 29 percent of days, and least likely around August 6, occurring in 23 percent of days. The most common forms of precipitation are light rain and thunderstorms. *Thunderstorms* happen during 74 percent of days with precipitation and are most likely around August 19 (21 percent). *Light rain* occurs during 15 percent of days with precipitation and are most likely around August 13 (5 percent).

Humidity

The relative humidity in August typically ranges from 23 percent to 90 percent, rarely dropping below 15 percent or reaching as high as 98 percent. The air is *driest* around August 9 (below 27 percent 3 days out of 4) and *most humid* around August 30 (above 85 percent 3 days out of 4).

Wind

Wind speeds vary from 0 mph to 15 mph, rarely exceeding 19 mph. The *highest* average wind speed occurs around August 3, and the *lowest* around August 17. The wind is most often out of the north (24 percent of the time), south (21 percent), and southwest (14 percent).

Riverton, Wyoming

Climate and Topography

Riverton has a cold semi-arid steppe climate. The area near this station is covered by *shrub lands* (71 percent), *forests* (13 percent), *grasslands* (11 percent), and *croplands* (3 percent).

Temperatures

August sees daily highs drop from 86 °F to 80 °F over the course of the month, exceeding 93 °F or dropping below 69 °F only 1 day in 10.

Cloud Cover

The median cloud cover is a marvelous 2 percent (clear). In mid-August, the sky is *clear or mostly clear* 64 percent of the time, *partly cloudy* 3 percent of the time, and *mostly cloudy or overcast* 15 percent of the time.

Rain

The average probability for rain on a given day is 25 percent, with little variation through the month. The most common forms of precipitation are light rain and thunderstorms, occurring during 57 and 29 percent of days with precipitation, respectively. Light rain is most likely around August 21 (15 percent). *Thunderstorms* are most likely at the beginning of the month (9 percent).

Humidity

The relative humidity typically ranges from 20 percent to 65 percent in August, rarely dropping below 11 percent or exceeding 90 percent. The air is *driest* around August 16 (below 24 percent 3 days out of 4) and *most humid* around August 12 (above 51 percent 3 days out of 4).

Wind

Typical wind speeds vary from 0 mph to 17 mph, rarely exceeding 27 mph. The *highest* average wind speed occurs around August 16 and the *lowest* around August 25. The wind is most often out of the west (12 percent of the time) and northwest (12 percent).

Casper, Wyoming

Climate and Topography

Casper has a humid continental climate with warm summers and no dry season. The area surrounding this station is covered by *shrub lands* (80 percent), *grasslands* (16 percent), and *forests* (4 percent).

Temperatures

August has daily highs that decrease from 87 °F to 82 °F over the course of the month, exceeding 94 °F or dropping below 72 °F only 1 day in 10.

Cloud Cover

The median cloud cover is 14 percent (mostly clear). In mid-August, the sky is *clear or mostly clear* 63 percent of the time, *partly cloudy* 15 percent of the time, and *mostly cloudy or overcast* 21 percent of the time.

Rain

The probability for rain varies throughout the month. It's most likely around August 1, occurring in 42 percent of days. Rain is least likely around August 31 (37 percent). The most common forms of precipitation are thunderstorms and light rain. *Thunderstorms* happen on 62 percent of days with precipitation and are most likely around August 1 (28 percent). *Light rain* occur during 29 percent of days with precipitation and are most likely around August 19 (12 percent).

Humidity

In August, the relative humidity typically ranges from 20 percent 77 percent, rarely dropping below 10 percent or reaching as high as 97 percent. The air is *driest* around August 26 (below 24 percent 3 days out of 4) and *most humid* around August 1 (above 67 percent 3 days out of 4).

Wind

Typical wind speeds vary from 1 mph to 19 mph, rarely exceeding 27 mph. The *highest* average wind speed occurs around August 30 and the *lowest* occurs around August 1. The wind is most often out of the southwest (38 percent of the time) and west (16 percent).

Carhenge (Based on Data from Alliance, Nebraska)

Climate and Topography

Alliance has a cold semi-arid steppe climate. The area around this station is covered by *grasslands* (97 percent).

Temperatures

August experiences *gradually falling* daily highs around 86 °F throughout the month, exceeding 97 °F or dropping below 72 °F only 1 day in 10.

Cloud Cover

The median cloud cover is 17 percent (mostly clear). In mid-August, the sky is *clear or mostly clear* 42 percent of the time, *partly cloudy* 12 percent of the time, and *mostly cloudy or overcast* 20 percent of the time.

Rain

The average probability for rain on a given day is 27 percent, with little variation over the course of the month. The most common forms of precipitation are thunderstorms and light rain. *Thunderstorms* happen during 55 percent of days with precipitation and are most likely around August 1 (17 percent). *Light rain* occurs on 33 percent of days with precipitation and is most likely around August 24 (9 percent).

Humidity

The relative humidity typically ranges from 30 percent to 97 percent in August, rarely dropping below 16 percent or reaching as high as 100 percent. The air is *driest* around August 30 (below 37 percent 3 days out of 4) and *most humid* around August 11 (above 96 percent 3 days out of 4).

Wind

Typical wind speeds vary from 1 mph to 18 mph, rarely exceeding 25 mph. The *highest* average wind speed occurs around August 8 and the *lowest* around August 1. The wind is most often out of the west (13 percent of the time) and south (11 percent).

Grand Island, Nebraska

Climate and Topography

Grand Island has a humid continental climate with hot summers and no dry season. The area near this station is covered by *croplands* (98 percent).

Temperatures

August has *gradually falling* daily highs around 85 °F throughout the month, exceeding 96 °F or dropping below 73 °F only 1 day in 10.

Cloud Cover

The median cloud cover is 16 percent (mostly clear). In mid-August, the sky is *clear or mostly clear* 60 percent of the time, *partly cloudy* 10 percent of the time, and *mostly cloudy or overcast* 26 percent of the time.

Rain

The probability of rain varies throughout the month. It is most likely around August 3, occurring in 43 percent of days and least likely around August 31 (38 percent). The most common forms of precipitation are thunderstorms (58 percent of days with precipitation), light rain (19 percent), and moderate rain (18 percent). *Thunderstorms* are most likely around August 1 (26 percent). *Light rain* is most likely around August 24 (8 percent). *Moderate rain* is most likely around August 15 (7 percent).

Humidity

The relative humidity in August typically ranges from 45 percent to 91 percent, rarely dropping below 29 percent or reaching as high as 99 percent. The air is *driest* around August 31 (below 52 percent 3 days out of 4) and *most humid* around August 15 (above 88 percent 3 days out of 4).

Wind

Typical wind speeds vary from 3 mph to 17 mph, rarely exceeding 23 mph. The *highest* average wind speed occurs around August 31 and the *lowest* around August 1. The wind is most often out of the south (24 percent of the time), north (17 percent), southeast (11 percent), northwest (11 percent), and west (10 percent).

St. Joseph, Missouri

Climate and Topography

St. Joseph has a humid continental climate with hot summers and no dry season. The area surrounding this station is covered by croplands (88 percent), grasslands (9 percent), and lakes and rivers (2 percent).

Temperatures

August exhibits *gradually falling* daily highs around 86 °F throughout the month, exceeding 95 °F or dropping below 77 °F only 1 day in 10.

Cloud Cover

The median cloud cover is a wonderful 4 percent (clear). In mid-August, the sky is *clear or mostly clear* 68 percent of the time, *partly cloudy* 4 percent of the time, and *mostly cloudy or overcast* 19 percent of the time. On August 11, the clearest day of the year, the sky is clear, mostly clear, or partly cloudy 73 percent of the time and mostly cloudy or overcast 18 percent of the time.

Rain

The average probability for rain on a given day is 36 percent, with little variation over the course of the month. The most common forms of precipitation are thunderstorms, light rain, heavy rain, and moderate rain. Thunderstorms occur on 52 percent of days with precipitation, light rain on 27 percent, heavy rain on 11 percent, and moderate rain on 10 percent. Thunderstorms are most likely around August 1

(21 percent of all days), light rain is most likely around August 21 (10 percent), heavy rain is most likely around August 24 (5 percent), and moderate rain is most likely around August 31 (4 percent).

Humidity

The relative humidity typically ranges from 50 percent to 94 percent in August, rarely dropping below 38 percent or reaching as high as 100 percent. The air is driest around August 31 (below 58 percent 3 days out of 4) and most humid around August 20 (above 91 percent 3 days out of 4).

Wind

Typical wind speeds vary from 0 mph to 13 mph, rarely exceeding 18 mph. The highest average wind speed occurs around August 3 and the lowest around August 30. The wind is most often out of the south (23 percent of the time) and north (16 percent).

Columbia, Missouri

Climate and Topography

Columbia has a humid continental climate with hot summers and no dry season. The area around this station is covered by *forests* (79 percent), *croplands* (9 percent), and *grasslands* (9 percent).

Temperatures

August has *gradually falling* daily highs around 86 °F throughout the month, exceeding 97 °F or dropping below 75 °F only 1 day in 10.

Cloud Cover

The median cloud cover is 21 percent (mostly clear). In mid-August, the sky is *clear or mostly clear* 60 percent of the time, *partly cloudy* 13 percent of the time, and *mostly cloudy or overcast* 26 percent of the time.

Rain

The average probability for rain on a given day is 36 percent, with little variation over the course of the month. The most common forms of precipitation are thunderstorms, light rain, and moderate rain. *Thunderstorms* happen on 59 percent of days with precipitation, are most likely around August 1, when they occur during 24 percent of all days. *Light rain* occurs on 21 percent of days with precipitation and is most likely around August 28 (9 percent). *Moderate rain* falls during 15 percent of days with precipitation and is most likely around August 27 (6 percent).

Humidity

The relative humidity in August typically ranges from 49 percent to 93 percent, rarely dropping below 33 percent or reaching as high as 100 percent. The air is *driest* around August 1 (below 58 percent 3 days out of 4) and *most humid* around August 16 (above 91 percent 3 days out of 4).

Wind

Typical wind speeds vary from 3 mph to 13 mph, rarely exceeding 18 mph. The *highest* average wind speed occurs around August 28 and the *lowest* occurs around August 2. The wind is most often out of the south (22 percent of the time), southeast (15 percent), northwest (12 percent), west (11 percent), east (11 percent), and north (10 percent).

Murphysboro, Illinois (Closest Weather Station to Carbondale)

Climate and Topography

Murphysboro, Illinois has a warm humid temperate climate with hot summers and no dry season. The area near this station is covered by *forests* (48 percent), *croplands* (46 percent), and *lakes and rivers* (3 percent).

Temperatures

August sees *gradually falling* daily highs around 87 °F throughout the month, exceeding 96 °F or dropping below 78 °F only 1 day in 10.

Cloud Cover

The median cloud cover for the month is 35 percent (mostly clear). In mid-August, the sky is *clear or mostly clear* 36 percent of the time, *partly cloudy* 13 percent of the time, and *mostly cloudy or overcast* 27 percent of the time.

Rain

The average probability for rain on a given day is 30 percent, with little variation over the course of the month. The most common forms of precipitation are thunderstorms, light rain, and moderate rain, with each occurring 61, 21, and 10 percent of days with precipitation respectively. Thunderstorms are most likely around August 2 (21 percent of all days). *Light rain* is most likely around August 31 (7 percent), and m*oderate rain* is most likely around August 23 (3 percent).

Humidity

The relative humidity typically ranges from 52 percent to 99 percent in August, rarely dropping below 35 percent or reaching as high as 100 percent. The air is *driest* around August 1 (below 58 percent 3 days out of 4) and *most humid* around August 7 (around 100 percent 3 days out of 4).

Wind

Typical wind speeds vary from 0 mph to 11 mph, rarely exceeding 16 mph. The *highest* average wind speed occurs around August 13 and the *lowest* occurs around August 4. The wind is most often out of the south (15 percent of the time).

Fort Campbell, Kentucky (Closest Weather Station to Hopkinsville)

Climate and Topography

Fort Campbell has a warm humid temperate climate with hot summers and no dry season. The area surrounding this station is covered by *forests* (85 percent), *croplands* (12 percent), and *lakes and rivers* (2 percent).

Temperatures

August daily highs are around 87 °F, exceeding 95 °F or dropping below 79 °F only 1 day in 10.

Cloud Cover

The median cloud cover is 57 percent (partly cloudy). In mid-August, the sky is *clear or mostly clear* 34 percent of the time, *partly cloudy* 22 percent of the time, and *mostly cloudy or overcast* 42 percent of the time.

Rain

The average probability of rain on a given day is 42 percent, with little variation over the course of the month. The most common forms of precipitation are thunderstorms (57 percent of days with precipitation) and light rain (27 percent). *Thunderstorms* are most likely at the start of the month (26 percent of all days). *Light rain* is most likely around month's end (12 percent).

Humidity

The relative humidity typically ranges from 50 percent to 93 percent in August, rarely dropping below 33 percent or reaching as high as 100 percent. The air is *driest* around August 29 (below 58 percent 3 days out of 4) and *most humid* around August 1 (above 88 percent 3 days out of 4).

Wind

Typical wind speeds vary from 0 mph to 10 mph, rarely exceeding 15 mph. The *highest* average wind speed occurs around August 25 and the *lowest* around August 3. The wind is most often out of the south (16 percent of the time), southwest (12 percent), and north (12 percent).

Nashville, Tennessee

Climate and Topography

Nashville has a warm humid temperate climate with hot summers and no dry season. The area around this station is covered by *forests* (84 percent), *built-up areas* (8 percent), *croplands* (4 percent), and *lakes and rivers* (3 percent).

Temperatures

August has *gradually falling* daily highs around 88 °F throughout the month, exceeding 96 °F or dropping below 79 °F only 1 day in 10.

Cloud Cover

The median cloud cover is 45 percent (partly cloudy). In mid-August, the sky is *clear or mostly clear* 43 percent of the time, *partly cloudy* 25 percent of the time, and *mostly cloudy or overcast* 32 percent of the time.

Rain

The average probability for rain on a given day is 36 percent, with little variation over the course of the month. The most common forms of precipitation are thunderstorms (56 percent of days with precipitation), light rain (24 percent), and moderate rain (14 percent). *Thunderstorms* are most likely around the beginning of the month (24 percent of all days). *Light rain* is most likely at month's end (10 percent), and m*oderate rain* is most likely around August 24 (6 percent).

Humidity

August's relative humidity typically ranges from 47 percent to 92 percent, rarely dropping below 32 percent or reaching as high as 99 percent. The air is *driest* around August 12 (below 54 percent 3 days out of 4) and *most humid* around August 1 (above 88 percent 3 days out of 4).

Wind

Typical wind speeds vary from 0 mph to 12 mph, rarely exceeding 16 mph. The *highest* average wind speed occurs around August 1 and the *lowest* around August 17. The wind is most often out of the south (26 percent of the time), north (15 percent), and southwest (10 percent).

Columbia, South Carolina

Climate and Topography

Columbia has a warm humid temperate climate with hot summers and no dry season. The area surrounding this station is covered by *croplands* (52 percent), *forests* (33 percent), *grasslands* (7 percent), *lakes and rivers* (4 percent), and *built-up areas* (4 percent).

Temperatures

August has *gradually falling* daily highs around 90 °F throughout the month, exceeding 97 °F or dropping below 82 °F only 1 day in 10.

Cloud Cover

The median cloud cover is 58 percent (partly cloudy). In mid-August, the sky is *clear or mostly clear* 39 percent of the time, *partly cloudy* 21 percent of the time, and *mostly cloudy or overcast* 39 percent of the time.

Rain

The probability for rain varies throughout the month. It's most likely around August 3 and least likely around month's end, occurring in equal amounts. The most common forms of precipitation are thunderstorms (57 percent of days with precipitation), moderate rain (27 percent), and light rain (12 percent). *Thunderstorms* are most likely at the start of August 1, when observers record them during 34 percent of all days. *Moderate rain* is most likely around the end of the month (15 percent), and *light rain* also is most likely around the end of the month (7 percent).

Humidity

In August, the relative humidity typically ranges from 49 percent to 95 percent, rarely dropping below 34 percent or reaching as high as 100 percent. The air is *driest* around August 1 (below 56 percent 3 days out of 4) and m*ost humid* around August 22 (above 93 percent 3 days out of 4).

Wind

Typical wind speeds vary from 0 mph to 12 mph, rarely exceeding 16 mph. The *highest* average wind speed occurs around August 29 and the lowest occurs around August 17. The wind is most often out of the west (15 percent of the time), southwest (15 percent), northeast (11 percent), and east (10 percent).

Chapter 25

20 Hot Spots for Viewing the Eclipse

It seems the two questions everyone wants answered are, "Where in the U.S. does totality happen?" and "Where are the best spots to view the event?" The Moon's umbra, after sweeping eastward across part of the Pacific Ocean, makes its initial landfall in Oregon. Then it gets interesting. Totality will pass through 11 more states before heading into the Atlantic Ocean to finish its sweep. The statistics and maps in this chapter will provide you with the necessary details of 20 of the best sites.

I do offer one suggestion when you're considering which of these spots to observe the eclipse from. Carefully consider the population of the place. A town of 10,000 is much more likely to have problems than a city of 75,000. Primary among the issues will be traffic. Small communities with one main road may suffer hours of gridlock. If you do choose to head to such a location, get there early, perhaps even a day or two ahead of the eclipse. Remember, August 21, 2017, is a Monday. So the whole weekend before will be free for most people.

Madras, Oregon

This town of some 6,500 residents lies 120 miles southeast of Portland, which itself sees a 99-percent partial eclipse. Close, but no cigar. Madras, however, probably boasts the best weather prospects of any location within the path of totality. In August, the average precipitation is less than 0.4 inch. I'm cautious about saying something like, "If you can get there, do it," because I'm not certain how many tourists the place can handle. Traffic on eclipse day, and even the weekend prior, may be brutal. Take a week-long vacation there, however, and you'll roll into the eclipse without a care in the world.

© Springer International Publishing Switzerland 2016

M.E. Bakich, *Your Guide to the 2017 Total Solar Eclipse*, The Patrick Moore Practical Astronomy Series, DOI 10.1007/978-3-319-27632-8_25

Fig. 25.1 The World War II Warbird Museum in Madras, Oregon, flies the largest U.S. flag in the state. Mount Jefferson stands in the background. (Courtesy of Joe Krenowicz)

Eclipse starts: 9:06:47 A.M. PDT
Eclipse ends: 11:41:04 A.M. PDT
Maximum eclipse: 10:20:38 A.M. PDT
Sun's altitude at maximum eclipse: 41.6°
Duration of totality: 2 minutes and 3 seconds
Width of the Moon's shadow: 62.6 miles (100.8 kilometers)

Lime, Oregon

This location may be my most outlandish suggestion, but hear me out. Lime is an unincorporated community that once had a thriving plant that produced lime. Big surprise, eh? Not many people stop by the abandoned town. For the eclipse, however, Lime has two advantages: great weather and easy access, right off Interstate 84.

Fig. 25.2 Lime, Oregon, used to have a thriving cement plant. For the eclipse, it could host a throng of people who would enjoy great weather. (Courtesy of Lyzadanger via Wikimedia Commons CC BY-SA 3.0)

Eclipse starts: 9:10:04 A.M. PDT
Eclipse ends: 11:47:55 A.M. PDT
Maximum eclipse: 10:25:54 A.M. PDT
Sun's altitude at maximum eclipse: 44.9°
Duration of totality: 2 minutes 10 seconds
Width of the Moon's shadow: 64 miles (102.9 kilometers)

Smiths Ferry, Idaho

According to the 2010 census, this burg had a population of 75, but if you're in western Idaho, this looks to be a great viewing site. Smiths Ferry sits at an elevation of 4,554 feet (1,388 meters) and lies along Idaho State Hwy. 55.

Fig. 25.3 Smiths Ferry is an unincorporated township in Idaho. (Map courtesy of Xavier M. Jubier; data courtesy of Google Imagery/TerraMetrics)

Eclipse starts: 10:11:14 A.M. MDT
Eclipse ends: 12:50:11 P.M. MDT
Maximum eclipse: 11:27:42 A.M. MDT
Sun's altitude at maximum eclipse: 45.9°
Duration of totality: 2 minutes and 11 seconds
Width of the Moon's shadow: 64.4 miles (103.6 kilometers)

Terreton, Idaho

This unincorporated community of 1,100 people sits at an elevation of 4,790 feet (1,460 meters). Terreton lies 35 miles northwest of Idaho Falls, and is a 2-hour drive to either West Yellowstone or Jackson Hole, Wyoming. It boasts great access from Interstate 15.

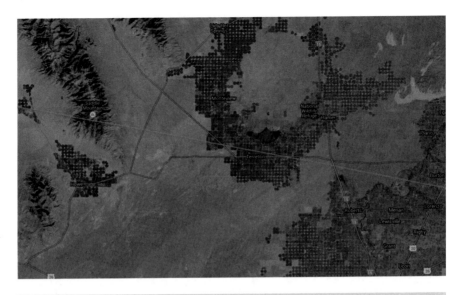

Fig. 25.4 Terreton is an unincorporated agricultural community in Idaho. (Map courtesy of Xavier M. Jubier; data courtesy of Google Imagery/TerraMetrics)

Eclipse starts: 10:14:57 A.M. MDT
Eclipse ends: 12:57:08 P.M. MDT
Maximum eclipse: 11:33:20 A.M. MDT
Sun's altitude at maximum eclipse: 49°
Duration of totality: 2 minutes and 17 seconds
Width of the Moon's shadow: 65.6 miles (105.5 kilometers)

Jackson Hole, Wyoming

Because of how scenic this location seems, lots of tour groups and individuals are headed here—in fact, perhaps too many for comfort. The town has 9,577 inhabitants, sits at an elevation of 6,311 feet (1,924 meters), and has a total area under 3 square miles. Jackson lies slightly south of the eclipse's center line, so if you drive north a few miles along U.S. Route 26, you'll experience an additional 4 seconds of totality. As demand and attendance will be high in this zone, if you want to experience totality from within this region, it is essential to plan early and book in advance.

Fig. 25.5 Looking west from Jackson Hole valley, the John Moulton Barn on Mormon Row stands at the base of the Grand Teton Mountains. (Courtesy of Wikipedia Creative Commons)

Eclipse starts: 10:16:43 A.M. MDT
Eclipse ends: 1:00:29 P.M. MDT
Maximum eclipse: 11:36:03 A.M. MDT
Sun's altitude at maximum eclipse: 50.5°
Duration of totality: 2 minutes and 15 seconds
Width of the Moon's shadow: 66.1 miles (106.3 kilometers)

Casper, Wyoming

Personal note: Wyoming's second-largest city had a population of 55,316 in 2010, and has increased by some 5,000 since. Its elevation is 5,200 feet (1,585 meters). Because Casper offers more amenities than most locations along the center line to its northwest, it will be the destination for vast numbers of visitors. Let's hope its 27 square miles of area can handle them.

Fig. 25.6 The skyline of Casper, Wyoming, is impressive. This city will host tens of thousands of eclipse watchers. (Courtesy of Wikipedia Creative Commons)

Eclipse starts: 10:22:17 A.M. MDT
Eclipse ends: 1:09:26 P.M. MDT
Maximum eclipse: 11:43:52 A.M. MDT
Sun's altitude at maximum eclipse: 54°
Duration of totality: 2 minutes and 26 seconds
Width of the Moon's shadow: 67.4 miles (108.5 kilometers)

Alliance, Nebraska

Were I not hosting a huge event in St. Joseph, Missouri, I would be in Alliance for the eclipse. Specifically, I'd head to Carhenge, a replica of Stonehenge built out of 39 automobiles. Great weather prospects, lots of open ground, and a modern (1987) construction that honors an ancient monument that marked the Sun's position. It doesn't get much better.

Fig. 25.7 The Carhenge monument lies three miles north of Alliance, Nebraska, along state highway 87. (Courtesy of Jacob Kamholz, via Wikimedia Commons CC BY-SA 4.0)

Eclipse starts: 10:27:09 A.M. MDT
Eclipse ends: 1:16:42 P.M. MDT
Maximum eclipse: 11:50:28 A.M. MDT
Sun's altitude at maximum eclipse: 56.7°
Duration of totality: 2 minutes 29 seconds
Width of the Moon's shadow: 68.4 miles (110 kilometers)

Stapleton, Nebraska

If you're looking for small-town charm, this berg with a population of 305 might just be for you. It lies at an elevation of 2,904 feet (885 meters), 29 miles north of Interstate 80 along U.S. Route 83, directly on the center line between the previous and the next entry on this list.

Fig. 25.8 This photo shows the east side of Main Street in Stapleton looking northeast from about Fifth Street. (Courtesy of Wikimedia Commons)

Eclipse starts: 11:30:46 A.M. CDT
Eclipse ends: 2:21:54 P.M. CDT
Maximum eclipse: 12:55:18 P.M. CDT
Sun's altitude at maximum eclipse: 58.5°
Duration of totality: 2 minutes and 33 seconds
Width of the Moon's shadow: 68.9 miles (110.9 kilometers)

Grand Island, Nebraska

Because of its population of some 50,000, this city can handle quite an influx of
eclipse watchers. With an elevation of 1,860 feet (567 meters), it's a bit lower than
more westerly locations. If you drive slightly south or southwest of the city to
where the center line intersects U.S. Route 281 or U.S. Route 30, respectively,
you'll pick up 1 extra second of totality.

Fig. 25.9 The Stuhr Museum of the Prairie Pioneer in Grand Island, Nebraska, preserves the
legacy of the Pioneers who settled the central Nebraska plains in the late 19th century. (Courtesy
of Wikimedia Commons)

Eclipse starts: 11:34:19 A.M. CDT
Eclipse ends: 2:26:35 P.M. CDT
Maximum eclipse: 12:59:50 P.M. CDT
Sun's altitude at maximum eclipse: 59.9°
Duration of totality: 2 minutes and 34 seconds
Width of the Moon's shadow: 69.4 miles (111.8 kilometers)

St. Joseph, Missouri (Rosecrans Memorial Airport)

Perhaps the largest single astronomy observing event ever will occur at Rosecrans Memorial Airport. If you're in the area, come join me there.

Fig. 25.10 The Missouri River flows through St. Joseph. (Courtesy of Tim Kiser, user Malepheasant, via Wikimedia Commons CC BY-SA 2.5)

Eclipse starts: 11:40:34 A.M. CDT
Eclipse ends: 2:34:28 P.M. CDT
Maximum eclipse: 1:07:39 P.M. CDT
Sun's altitude at maximum eclipse: 61.9°
Duration of totality: 2 minutes and 38 seconds
Width of the Moon's shadow: 70.1 miles (112.9 kilometers)

Columbia, Missouri (UM Stadium)

Another huge event will occur in the Show-Me State when astronomer Angela
Speck hosts the public at Faurot Field at Memorial Stadium, where the University
of Missouri football team plays. This facility has a capacity of 75,982.

Fig. 25.11 Faurot Field at Memorial Stadium on the University of Missouri campus in
Columbia will host some 80,000 eclipse watchers August 21, 2017. (Courtesy of Clare Murphy/
KOMU, Flickr user KOMUnews, CC BY 2.0)

Eclipse starts: 11:45:41 A.M. CDT
Eclipse ends: 2:40:15 P.M. CDT
Maximum eclipse: 1:13:41 P.M. CDT
Sun's altitude at maximum eclipse: 62.9°
Duration of totality: 2 minutes and 37 seconds
Width of the Moon's shadow: 70.6 miles (113.6 kilometers)

St. Clair, Missouri

St. Louis lies on the northeastern edge of the eclipse path, so it won't offer much in the way of length of totality. Instead, head southwest via Interstate 44 to the center line in St. Clair, a town of some 5,000 residents.

Fig. 25.12 These water towers outside Saint Clair, Missouri, have the labels "Hot" and "Cold" painted on them. (Courtesy of James Hayes from USA. Uploaded by LongLiveRock) via Wikimedia Commons CC BY 2.0)

Eclipse starts: 11:48:33 A.M. CDT
Eclipse ends: 2:43:29 P.M. CDT
Maximum eclipse: 1:17:04 P.M. CDT
Sun's altitude at maximum eclipse: 63.4°
Duration of totality: 2 minutes and 40 seconds
Width of the Moon's shadow: 70.8 miles (114 kilometers)

Carbondale, Illinois

Although this city of nearly 30,000 lies slightly off the center line, I offer it as a
choice because it's much larger than the other towns around it that are on the center
line. As I write this, big eclipse plans are underway at Southern Illinois University,
headed by physics and astronomy professor Bob Baer, making this a great choice
for viewing.

Fig. 25.13 This statue in downtown Carbondale salutes its railroaders and the fact that the city
was a thriving train gateway, with the first train passing through July 4, 1854. (Courtesy of
Explorecdale/Wikimedia Commons)

Eclipse starts: 11:52:27 A.M. CDT
Eclipse ends: 2:47:30 P.M. CDT
Maximum eclipse: 1:21:26 P.M. CDT
Sun's altitude at maximum eclipse: 63.7°
Duration of totality: 2 minutes 38 seconds
Width of the Moon's shadow: 71.1 miles (114.4 kilometers)

Giant City State Park, Makanda, Illinois

I hate to be the bearer of both good and bad news, but this attraction has the longest duration of totality anywhere. That's the good news. The other side of that coin is that the town of Makanda has a population of 561, and thousands of unknowing eclipse-watchers, hopeful of being in the location that offers the longest duration of totality, surely will flock there. Repeat after me: "Gridlock!"

Fig. 29.14 Technically, the longest duration of the August 21, 2017, total solar eclipse will occur at Giant City State Park in Illinois. (Courtesy of Alan Scott Walker via Wikimedia Commons)

Eclipse starts: 11:52:33 A.M. CDT
Eclipse ends: 2:47:43 P.M. CDT
Maximum eclipse: 1:21:37 P.M. CDT
Sun's altitude at maximum eclipse: 63.8°
Duration of totality: 2 minutes and 40 seconds
Width of the Moon's shadow: 71 miles (114.3 kilometers)

Cerulean, Kentucky

This unincorporated community has a population of about 400, it's easy to reach, it
has essentially the longest duration of totality, and, most importantly, many people
may overlook it as a prime site, making it a good area to target for viewing. Note
that if you plan to observe totality anywhere in Kentucky, you'll be in the Central
Time Zone. Most of the state observes Eastern Time, but those locations will see
only a partial eclipse.

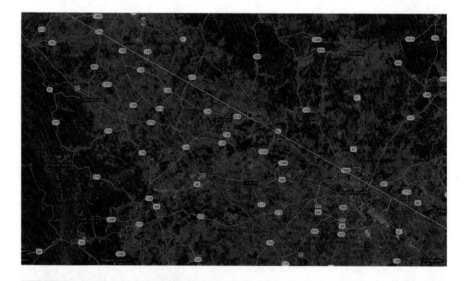

Fig. 25.15 Cerulean is an unincorporated community in Trigg County, Kentucky. (Map courtesy
of Xavier M. Jubier; data courtesy of Google Imagery/TerraMetrics)

Eclipse starts: 11:56:01 A.M. CDT
Eclipse ends: 2:51:12 P.M. CDT
Maximum eclipse: 1:25:27 P.M. CDT
Sun's altitude at maximum eclipse: 63.9°
Duration of totality: 2 minutes and 40 seconds
Width of the Moon's shadow: 71.2 miles (114.6 kilometers)

Hopkinsville, Kentucky

Because of a duration of totality near the maximum visible anywhere and an aggressive ad campaign produced by astronomy enthusiasts, this city of nearly 35,000 residents had better prepare itself for an influx of observers. It does boast easy access: Just take U.S. Route 68 roughly 17 miles east from where it meets Interstate 24.

Fig. 25.16 The First Presbyterian Church of Hopkinsville stands on land purchased for it in 1822. (Courtesy of Wikimedia Commons)

Eclipse starts: 11:56:33 A.M. CDT
Eclipse ends: 2:51:43 P.M. CDT
Maximum eclipse: 1:26:02 P.M. CDT
Sun's altitude at maximum eclipse: 63.9°
Duration of totality: 2 minutes and 40 seconds
Width of the Moon's shadow: 71.2 miles (114.6 kilometers)

Gallatin, Tennessee

Although Nashville is the single largest city that lies within the path of totality, I recommend you take U.S. Route 31E northeast 29 miles to Gallatin instead. Depending on where you start in Nashville, you'll gain up to nearly 1 minute of totality by doing so. For an extra minute of totality, I would even walk the 29 miles. So would just about everyone I know who has experienced a total solar eclipse—the extra time is that worth it!

Fig. 25.17 Downtown Gallatin is a quaint space. On eclipse day, however, expect it to be bustling with activity. (Courtesy of Ichabod/Wikimedia Commons)

Eclipse starts: 11:59:04 A.M. CDT
Eclipse ends: 2:54:11 P.M. CDT
Maximum eclipse: 1:28:47 P.M. CDT
Sun's altitude at maximum eclipse: 63.9°
Duration of totality: 2 minutes and 40 seconds
Width of the Moon's shadow: 71.3 miles (114.7 kilometers)

Sparta, Tennessee

This town of 5,100 people should prove a relatively quiet setting from which to observe the eclipse. If you like regional history and gorgeous buildings, this might just be your spot.

Fig. 25.18 The Rock House in Sparta, Tennessee, was built in the late 1830s and served as a stage coach inn. The guardrail of US-70 spans the embankment in the background. (Courtesy of Brian Stansberry via Wikimedia Commons)

Eclipse starts: 12:01:31 P.M. CDT
Eclipse ends: 2:56:29 P.M. CDT
Maximum eclipse: 1:31:24
Sun's altitude at maximum eclipse: 63.7°
Duration of totality: 2 minutes and 39 seconds
Width of the Moon's shadow: 71.4 miles (114.8 kilometers)

Greenville, South Carolina

While this city of more than 62,000 residents lies within the path of the Moon's
shadow, if you head 28 miles southwest to anywhere near the intersection of State
Route 81 and Interstate 85, you'll gain nearly half a minute of totality. That's the
spot the following statistics are for.

Fig. 25.19 The Falls in downtown Greenville are often visited by locals and tourists alike.
They'll be much more difficult to see when the Moon's shadow envelops them. (Courtesy of
Yousef Abdul-Husain via Wikimedia Commons)

Eclipse starts: 1:08:59 P.M. EDT
Eclipse ends: 4:03:01 P.M. EDT
Maximum eclipse: 2:39:03 P.M. EDT
Sun's altitude at maximum eclipse: 62.8°
Duration of totality: 2 minutes and 37 seconds
Width of the Moon's shadow: 71.5 miles (115 kilometers)

Columbia, South Carolina

If the sky is clear, this city of some 135,000 is sure to be the destination of tens to hundreds of thousands of hopeful eclipse chasers from locations up and down the Eastern Seaboard. And it might be one of the few spots able to handle such numbers.

Fig. 25.20 Columbia is the third-largest city covered by the Moon's umbra. Officials are bracing for the influx of several hundred thousand people. (Courtesy of Akhenaton06 via Wikimedia Commons)

Eclipse starts: 1:13:08 P.M. EDT
Eclipse ends: 4:06:21 P.M. EDT
Maximum eclipse: 2:43:07 P.M. EDT
Sun's altitude at maximum eclipse: 61.9°
Duration of totality: 2 minutes and 30 seconds
Width of the Moon's shadow: 71.5 miles (115 kilometers)

Appendix A

Resource list

Astronomy magazine (www.Astronomy.com)

Astronomy offers you the most exciting, visually stunning, thorough, and timely coverage of the heavens above. Each monthly issue includes expert science reporting, vivid color photography, complete sky-event coverage, spot-on observing tips, informative telescope reviews, and more. All of this comes in an easy-to-understand user-friendly style that's perfect for astronomers at any level. And in the lead-up to the eclipse in 2017, *Astronomy* will publish stories about the eclipse in the magazine, plus news releases and blogs on the website—most of them authored by yours truly.

Celestron (www.celestron.com)

The giant of the telescope industry has been selling telescopes, binoculars, and accessories for more than half a century. Most telescope dealers throughout the U.S. carry Celestron products.

Eclipser (www.eclipser.ca)

Retired Canadian meteorologist Jay Anderson operates this site. Astronomers and eclipse-chasers have been relying on his predictions for two decades. One nice feature here is that you can find climatology for future eclipses early, several years in advance, in fact. But Anderson continually revises his text as the event approaches.

© Springer International Publishing Switzerland 2016
M.E. Bakich, *Your Guide to the 2017 Total Solar Eclipse*, The Patrick Moore
Practical Astronomy Series, DOI 10.1007/978-3-319-27632-8

Front Page Science (www.stjosepheclipse.com)

Here you'll find all the information you need about the huge, free public eclipse-watching event at Rosecrans Memorial Airport in St. Joseph, Missouri.

The Great American Eclipse (www.greatamericaneclipse.com)

This site, run by longtime mapmaker and eclipse chaser Michael Zeiler, has the coolest graphics anywhere related to eclipses, and particularly the August 21, 2017 event. You'll find a variety of U.S. maps, incredibly detailed individual state maps, and even a 10-foot-long map of the eclipse path, which you can print for personal use.

iTunes (Search: 2017 Total Solar Eclipse)

This podcast, begun in 2015 by the author, contains all relevant information about the August 21, 2017 event in an easy-to-digest audio format. Most of the podcasts are in the 5–15 minute range, making them perfect for short commutes to work.

Mr. Eclipse (www.MrEclipse.com)

Longtime NASA employee (now retired) Fred Espenak runs this site, and you should consider it your "one-stop shopping" site for eclipse information. So, take this book (which I have tried to make quite thorough), multiply its information by, oh, 100, and you'll get some idea as to what you'll find on MrEclipse.com.

NASA (eclipse.gsfc.nasa.gov)

It's hard to beat the government in anything. That goes for eclipse sites, too, especially in the crucial few days leading up to the event. Other sites may be overloaded by the volume of traffic, but NASA's servers are spectacularly robust. This site also hosts a historical archive where you'll find information about eclipses going back 5 millennia.

Rainbow Symphony (www.rainbowsymphony.com)

Don't go to this site for eclipse information. Rather, head there to find a wide variety of approved solar viewing glasses and telescope and binocular filters. You'll find designer wrap-around glasses for $20 or (equally safe) cardboard glasses for much less than a dollar in quantity.

TravelQuest (www.travelquesttours.com)

This company is one of the world's most successful astronomical tour companies. Whether chasing the aurora borealis or sharing the spectacle of a total solar eclipse, TQ takes its clients to the best spots on the planet to experience astronomical wonders—from the North Pole to the South Pole and across Asia, Africa, Europe and the Americas. Travelers are from all walks of life. And while TQ caters primarily to a North American clientele, the company regularly welcomes people from every corner of the globe.

Appendix B

A Calendar of Every Total Solar Eclipse Since the Year 1

Some statistics are confusing, but others are truly cool. To demonstrate that total solar eclipses can occur on any date, here is a calendar in which each date box contains the year(s) of any total or hybrid solar eclipse that has occurred on that date from the year 1 through the great event in 2017. Have any eclipses happened on your birth date? There's no significance to it, but it's cool to know nonetheless?

January

Sunday	Monday	Tuesday	Wednesday	Thursday	Friday	Saturday
1 0865 1386 1405 1889	2 0447 0466 0987	3 1535 1554 1908 1573	4 0493 1014 1033	5 0075 1685	6 0102 0121 0642 0661 1163 1182	7 1209
8 1228 1712 1731 1750	9 0270 0688	10 0297 0837 1358 1377	11 0316 0856 1880	12 0883 1404 1423	13 0465 0986 1005	14 0484 1553 1907 1926
15 0511 1032 1051 1572	16 0093	17 1181 1200 1703	18 0120 0139 0660 0679 1730 1749	19 0706 1227 1246 1768	20 0288	21 1376 1395
22 0315 0334 0836 0855 0874 1898	23 0901 1422 1441	24 0483 0502 1004 1023 1544 1925	25 1050 1069 1571 1591 1944	26 0529	27 0111 1721	28 0138 0157 0659 0678 0697 1199 1218
29 1245	30 0724 1264 1748 1767 1786	31 0306				

Fig. B.1

© Springer International Publishing Switzerland 2016
M.E. Bakich, *Your Guide to the 2017 Total Solar Eclipse*, The Patrick Moore Practical Astronomy Series, DOI 10.1007/978-3-319-27632-8

February

Sunday	Monday	Tuesday	Wednesday	Thursday	Friday	Saturday
			1 0333 0854 0873 1394 1413	2 0352 0892	3 0919 1041 1440 1562 1916 1022 1459	4 0501 1590 1943 1609
5 0520 1962	6 0129 0547 1068 1087	7 0677 1217	8 0696 0715 1236 1739	9 0156 0175 1766 1785	10 0742 0845 1263 1282	11 0324 1804
12 0351 0370 0872 0891 0910 1412 1431	13 1458 1477	14 1934	15 0500 0519 0538 1040 1059 1580 1589 1961	16 0565 1086 1105 1608 1627 1980	17	18 0147 1757
19 0174 0193 0695 0714 0733 1235 1254	20 1281 1784	21 0342 0361 0760 0863 1300 1803	22 0323 1430 1449	23 0890 0909	24 0388 0928	25 0518 0537 1058 1077 1476 1495 1952
26 0016 0035 0556 1607 1626 1645 1979 1998	27 1104 1123	28 0165 0583 0686	29			

Fig. B.2

March

Sunday	Monday	Tuesday	Wednesday	Thursday	Friday	Saturday
			1 0192 0713 0732 1253 1775	2 0211 0751	3 0778 1299 1318	4 0341 0360 0881 1802 1821
5 0379 1448	6 0406 0927 0946 1467	7 1076 1494 1513 1598 1970	8 0536 1095 1625 1644	9 0015 0034 0053 0555 0574 1663 1997 2016	10 0164 0601 0704 1122 1141	11 0183
12 0731 0750 0769 1271 1290 1793	13 0210 0229	14 0796 1317 1336 1820	15 0359 0378 0397 0899 0918 1839	16 0424 0945 0964 1466 1485	17 1616	18 0006 0527 1531 1988
19 0033 0052 0554 0573 0592 1094 1113	20 0071 1140 1643 1662 1681 2015	21 0619 0722 1159	22 0182 0201	23 0228 0749 0768 1289 1308	24 0247 0787 1811	25 0377 0396 0814 0917 0936 1335 1354 1838 1857
26 0415 1484	27 1503 1522	28 0005 0024 0442 0545 0963 0982	29 0572 1112 1549 1634 1987 2006	30 0051 0070 0089 0591 0610 1131 1661 1680	31 1158 1177 1699	

Fig. B.3

April

Sunday	Monday	Tuesday	Wednesday	Thursday	Friday	Saturday
						1 0200 0637 0740
2 0219	**3** 0246 0265 0767 0786 0805 1307 1326 1829	**4** 0832 1353 1372	**5** 1475 1856	**6** 0395 0414 0433 0935 0954 1875 1894	**7** 0460 0981 1000 1502 1521 1540	**8** 0042 0563 1652 2005
9 0023 1130 1149 1567	**10** 0069 0088 0590 0609 0628 1679 1698	**11** 0107 1176 1717	**12** 0218 0237 0256 0655 0758 1195	**13** 0785 0804 1325 1344	**14** 0823 1828	**15** 0283 0386 1390 1847
16 0413 0432 0953 0972 1371 1493 1874 1893	**17** 0451 1520 1912	**18** 0478 0999 1018 1539 1558	**19** 0041 0060 0581 1670	**20** 0608 1148	**21** 0087 0106 0125 0627 0646 1167 1697	**22** 0673 0776 1194 1213 1716
23 0236 0795 1735	**24** 0255 0274	**25** 0301 0404 0822 0841 1343 1362 1846 1865	**26** 1408 1892	**27** 0431 0450 0469 0971 0990 1511	**28** 1538 1557 1576 1911 1930	**29** 0496 1017 1036 1585
30 0059 0078 0599 1669 1688						

Fig. B.4

May

Sunday	Monday	Tuesday	Wednesday	Thursday	Friday	Saturday
	1 0105 0124 0626 0645 0664 1166 1185	**2** 0143 1212	**3** 0227 0691 1231 1715 1734 1753	**4** 0254 0273 0292 0794 0813	**5** 0840 1361 1380	**6** 0319 0422 0859 1864 1883
7 1426 1529	**8** 0449 0468 0989 1008 1510	**9** 0487 1556 1910 1929	**10** 0077 0096 0514 0617 1035 1054 1575	**11** 0115 1184 1687	**12** 0644 1203 1706	**13** 0142 0161 0245 0663 0682 1733 1752
14 0272 0812 1230 1249 1352 1771	**15** 0291 0310 0831	**16** 0337 0858 0877 1379 1398 1417	**17** 0440 1444 1882	**18** 1528 1901	**19** 0467 0486 0505 1007 1026 1547 1928	**20** 0532 1053 1072 1574 1594 1947
21	**22** 0095 0114 0133 0635 1705 1724	**23** 0160 0681 0700 1202 1221	**24** 0179 0263 1248 1789	**25** 0309 0328 0830 0849 1267 1351 1370 1751 1770	**26** 0290 1397	**27** 0876 1416 1435
28 0355 0458 0895 1900	**29** 0485 0504 1025 1044 1462 1546 1565 1919 1938	**30** 0002 0523 1593 1612 1965	**31** 0086 1071 1090			

Fig. B.5

June

Sunday	Monday	Tuesday	Wednesday	Thursday	Friday	Saturday
				1 0113 0132 0550 0653 0672	2 0151 1220	3 0178 0699 0718 1239 1723 1742
4 0281 1285 1388 1769 1788	5 0308 0848 1369	6 0327 0346 0867 1807	7 0373 0476 0894 0913 1415 1434 1453	8 1480 1564 1918 1937 1956	9 1043 1062 1592	10 0001 0020 0104 0503 0522 0541 1611 1630
11 1089 1108 1983	12 0131 0150 0169 0671 0690 1211	13 0717 0736 1238 1257 1276 1741 1760	14 0196	15 0299 1303 1787	16 0326 0345 0364 0866 0885 1406 1806 1825	17 0912 1433 1452 1909
18 0391 0931 1471	19 0494 1583 1936	20 0540 1061 1080 1582 1955 1974	21 0019 0038 0559 1610 1629 1648 2001	22 0689 0708 1107 1126 1210 1229	23 0149 0168	24 0187 1256 1759 1778 1797
25 0214 0317 0735 0754 1275 1294	26 0344 0884 1321 1405 1424 1824	27 0363 0382 0903 1843	28 1451 1470 1489	29 0409 0512 0930 0949 1927	30 0010 0531 1601 1954 1973 1992	

Fig. B.6

July

Sunday	Monday	Tuesday	Wednesday	Thursday	Friday	Saturday
						1 0037 0558 0577 1079 1098 1628
2 0140 1144 1647 1666	3 1228	4 0167 0186 0205 0707 0726 1247 1796	5 0232 0753 0772 1274 1293 1312 1777	6 0335 1815	7 1339 1442	8 0362 0381 0400 0902 0921 1423 1842
9 0948 1469 1488 1945	10 0009 0028 0427 0530 0549 0967 1070 1507 1600 1972	11 1097 1116 1619 1991 2010	12 0576 1135 1646 1665 1684	13 0055 0158 0595	14 0185 0204 0725 0744 1162 1246 1265 1768	15 0223 1292
16 0771 0790 1311 1330 1795	17 0250 0353 1814 1833	18 0920 1441 1460 1860	19 0399 0418 0939	20 0966 0985 1088 1487 1506 1525 1963	21 0008 0548 1618 1637 1656	22 0027 0046 0567 1990 2009
23 0073 0176 0594 0613 1115 1134 1153	24 1180 1264 1683 1702	25 0743 0762 1283 1786	26 0203 0222 0241 1348	27 0268 0789 0808 1310 1329 1813 1832	28 0371 0390 1851	29 0417 0436 0938 0957 1459 1478 1878
30 1505 1524	31 1003 1543 1981					

Fig. B.7

August

Sunday	Monday	Tuesday	Wednesday	Thursday	Friday	Saturday
		1 0026 0045 0064 0566 0585 1106 1636 2008	2 0612 1133 1152 1655 1674	3 0091 0631 1171	4 0194 1701	5 0221 0240 0761 0780 1282 1301 1804
6 0259 1328	7 0286 0389 0408 0807 0826 1347 1366 1831 1850 1869	8 1477 1496	9 0956 0994 1896	10 0017 0435 0975	11 0044 0584 1021 1105 1124 1523 1542 1627 1999	12 0063 0082 0603 1654 1673 1692
13 0649 1151 1170 1189	14 0109 0212 0630	15 1300 1719	16 0239 0258 0277 0779 0798 1319 1822	17 0825 0844 1346 1365	18 1849 1868	19 0407 0426 0947 1887
20 0472 0974 0993 1012 1495 1514 1533	21 0035 1560 1645 1914 2017	22 1123 1142 1672	23 0062 0081 0100 0602 0621 1691	24 0648 1169 1188 1710	25 0127 0230 0667 1207	26 0797 0816 1318 1337 1737
27 0276 0295 1840	28 1364	29 0425 0843 0862 0965 1383 1867 1886	30 0992 1513 1905	31 0471 0490 1011 1030 1551 1932		

Fig. B.8

September

Sunday	Monday	Tuesday	Wednesday	Thursday	Friday	Saturday
					1 0053 1578 1663	2 0080 0620 1141 1160
3 0099 0118 0639 1690	4 0145 0666 0685 1187 1206 1709 1728	5 0248	6 0267 1336 1755	7 0294 0313 0815 0834 1355 1858	8 0861 1382 1401 1885	9 0443 0462 0983 1904
10 1010 1531 1550 1569 1923	11 0489 0508 1029	12 0071 1205 1681 1950	13 0098 0117 0136 0638 0657 1159 1178	14 1224 1708	15 0684 0703 1727 1746	16 0163 0266 0285 1773
17 0312 0833 0852 1354 1373 1876	18 0331 1400	19 1419	20 0461 0480 1001	21 1028 1549 1568 1903 1922 1941	22 0089 0507 0526 1047 1596 1968	23 1177 1196 1699
24 0116 0656	25 0135 0154 0675 1726 1745 1764	26 0284 0702 1223 1242	27 0303 1791	28 0349 0851 0870 1391	29 0330 1894	30 1418 1437

Fig. B.9

October

Sunday	Monday	Tuesday	Wednesday	Thursday	Friday	Saturday
1 0479 0498 1019 1921 1940	2 0004 0525 0544 1046 1065 1567 1587 1959	3 1614 1986	4 0107 1717	5 0134 0153 0172 0674 0693 1195 1214	6 0720 1241 1260 1744	7 0842 1763 1782
8 0302 0321	9 0348 0869 0888 1390 1409 1428 1809	10 0367 1912	11 0497 0516 1037 1455	12 1586 1605 1939 1958 1977	13 1064 1632	14 0022 0125 0543 1083
15 0152 0692 1213 1232	16 0171 0190 0711 1735	17 1259 1278 1762 1781	18 0320 0738 0860 1800	19 0339 1408	20 0366 0385 0887 0906 1427 1827	21 1473 1930
22 1585 1604	23 0013 0515 0534 1055 1623 1957 1976	24 0561 1082 1101 1995	25 0143 1650	26 1250 1753	27 0170 0189 0208 0710 0729 1780	28 0756 1277 1296 1799
29 0338 0357 0878 1818	30 0905 0924 1426 1445 1464 1845	31 0384				

Fig. B.10

November

Sunday	Monday	Tuesday	Wednesday	Thursday	Friday	Saturday
			1 0403 1948	2 0012 0533 0552 1073 1491 1967	3 0031 1100 1603 1622 1641 1994 2013	4 0579 1119 1668
5 0161	6 0188 0728 1249 1268 1771	7 0207 0226 0747	8 0774 0896 1295 1314 1798	9 0356 1817 1836	10 0375 1444	11 0402 0923 0942 1463 1482
12 1509 1966 1985	13 0551 0570 1621 1640 2012	14 0011 0030 0049 1659	15 0597 1118 1137 1686	16 0179	17 0206 0225 0244 0746 0765 1267 1286 1789	18 1313 1332
19 0792 1816	20 0374 0393 0914 1835 1854	21 0420 0941 0960 1462 1481 1500	22 1984	23 0002 1527 2003	24 0029 0048 0569 0588 1109 1639 1658 1677	25 0067 1136
26 0615 1155	27 0197 1304	28 0224 0764 1285	29 0243 1807	30 0392 0810 0932 1331 1350 1834 1853 1872		

Fig. B.11 V

December

Sunday	Monday	Tuesday	Wednesday	Thursday	Friday	Saturday
					1 0411 1480	2 1499 1518
3 0020 0438 0959 0978	4 1545 1630 2002	5 0047 0066 0085 0587 0606 1127 1657 1676	6 1154 1173 1695	7 0633	8 0215	9 0242 0261 0782 0801 1303 1322 1825
10 0828 1349 1368	11 1852	12 0410 0429 0950 1871 1890	13 0456 0977 0996 1498 1517 1536	14 0038 1648	15 1126 1145 1563	16 0065 0084 0605 0624 1675 1694
17 0103 1172 1713	18 0233 0651 1191	19 0800 1321 1340	20 0260 0819	21 0279 1843	22 0428 0846 0968 1367 1386 1870 1889	23 0447 1516 1908
24 0474 0995 1014 1535 1554	25 0056 1666	26 1144	27 0083 0102 0121 0623 0642 1163 1693	28 0669 1190 1209 1712	29 1731	30 0251
31 0278 0297 0818 0837 0358 1339 1861						

Fig. B.12

Appendix C

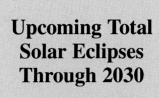

**Upcoming Total
Solar Eclipses
Through 2030**

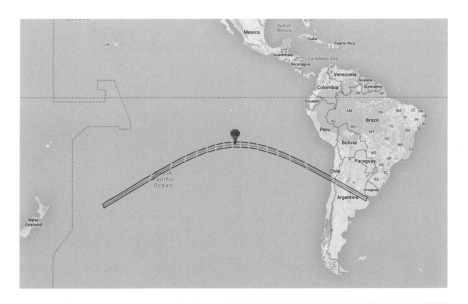

Fig. C.1 July 2, 2019
The maximum duration of totality, 4 minutes and 33 seconds, occurs at 19h24m07s Universal
Time (11:24:07 P.M. EDT). The Moon's shadow during this eclipse falls primarily over the
South Pacific Ocean. Only two countries, Chile and Argentina will host landlocked observers.
Xavier M. Jubier; data: INEGI Imagery/NASA/TerraMetrics

© Springer International Publishing Switzerland 2016   321
M.E. Bakich, *Your Guide to the 2017 Total Solar Eclipse*, The Patrick Moore
Practical Astronomy Series, DOI 10.1007/978-3-319-27632-8

Fig. C.2 December 14, 2020
The maximum duration of totality, 2 minutes and 10 seconds, occurs at 16h14m39s Universal Time (9:14:39 P.M. EST). As with the total solar eclipse in the previous year, this one begins in the South Pacific Ocean and hits land only in Chile and Argentina before moving into the South Atlantic Ocean. Xavier M. Jubier; data: INEGI Imagery/NASA/TerraMetrics

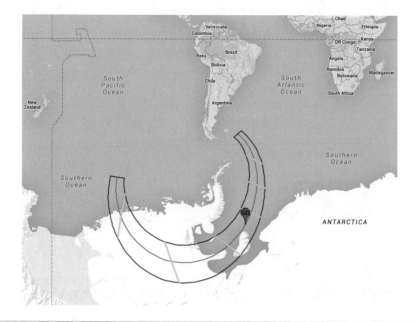

Fig. C.3 December 4, 2021
The maximum duration of totality, 1 minute and 54 seconds, occurs at 7h34m38s Universal Time (12:34:38 P.M. EST). This eclipse will be somewhat hard to observe. While a tiny bit of totality falls over the southern Atlantic Ocean, most of the Moon's shadow darkens either the Southern Ocean or Antarctica. Xavier M. Jubier; data: INEGI Imagery/NASA/TerraMetrics

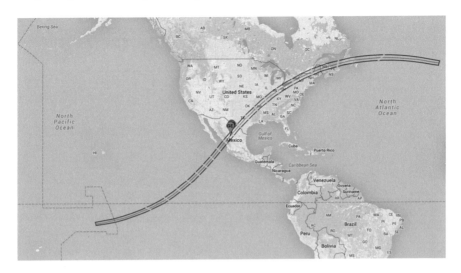

Fig. C.4 April 8, 2024
The maximum duration of totality, 4 minutes and 28 seconds, occurs at 18h18m29s Universal Time (10:18:29 P.M. EDT). This is the next great total solar eclipse for the U.S. And while most of the path lies over the Pacific and Atlantic Oceans, vast numbers of land-based observers in Mexico, the U.S., and tiny parts of eastern Canada will be able to view totality. Xavier M. Jubier; data: INEGI Imagery/NASA/TerraMetrics

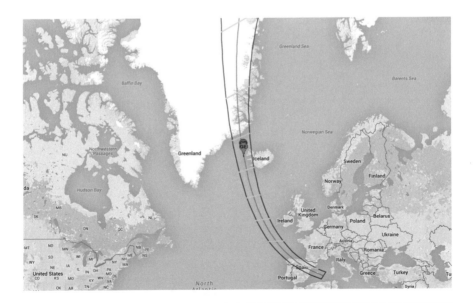

Fig. C.5 August 12, 2026
The maximum duration of totality, 2 minutes and 18 seconds, occurs at 17h47m05s Universal Time. The best viewing regions will be the Arctic, Greenland, Iceland, and Spain. Xavier M. Jubier; data: INEGI Imagery/NASA/TerraMetrics

Fig. C.6 August 2, 2027
The maximum duration of totality, a whopping 6 minutes and 23 seconds, occurs at 10h07m49s Universal Time. Observers under clear skies in Morocco, Spain, Algeria, Libya, Egypt, Saudi Arabia, Yemen, and Somalia will have the best opportunities to experience this ultra-long totality. Xavier M. Jubier; data: INEGI Imagery/NASA/TerraMetrics

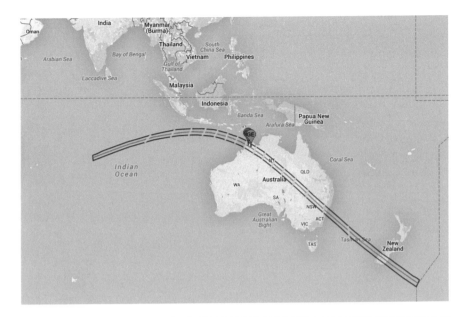

Fig. C.7 July 22, 2028
The maximum duration of totality, a worthy 5 minutes and 10 seconds, occurs at 2h56m39s
Universal Time. The best regions to view this event will be in Australia and New Zealand.
Xavier M. Jubier; data: INEGI Imagery/NASA/TerraMetrics

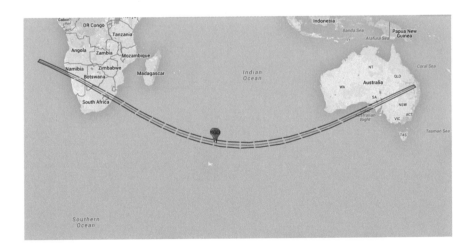

Fig. C.8 November 25, 2030
The maximum duration of totality, 3 minutes and 44 seconds, occurs at 6h51m37s Universal
Time. Viewers under clear skies along the path of totality in Botswana, South Africa, and
Australia will see this event. Xavier M. Jubier; data: INEGI Imagery/NASA/TerraMetrics

Appendix D

Eclipse-Related Timeline 2016–2017

What are the benchmark dates from the publication of this book until the eclipse?

Date	Event
March 9, 2016	A total solar eclipse occurs over Indonesia
April 8, 2016	Only 500 days until the 2017 eclipse
April 17–23, 2016	The Sun's path is the same as on eclipse day
August 21, 2016	Just 1 year until the 2017 eclipse
September 1, 2016	An annular eclipse occurs over central Africa
December 14, 2016	A scant 250 days are left until the 2017 eclipse
February 26, 2017	An annular eclipse occurs over South America and Africa
April 17–23, 2017	The Sun's path is the same as on eclipse day
May 13, 2017	Barely 100 days remain until the 2017 eclipse
August 7, 2017	A partial eclipse of the Moon (24.6 percent) is visible from Europe, Africa, and Asia
August 21, 2017	THE BIG DAY!

© Springer International Publishing Switzerland 2016   327
M.E. Bakich, *Your Guide to the 2017 Total Solar Eclipse*, The Patrick Moore
Practical Astronomy Series, DOI 10.1007/978-3-319-27632-8

Appendix E

Telescopes 101

Buying a telescope is a big step, especially if you're not sure what all those terms—f-ratio, magnification, go-to—mean. To eliminate confusion and make sure you understand what you're buying, here are 11 things to check out before you write the check out.

1. I know telescopes make things appear bigger, but what *exactly* do they do?

A telescope's main purpose is to collect light. This property of telescopes allows you to observe objects much fainter than you can see with your eyes alone. Galileo said it best when he declared that his telescopes "revealed the invisible." And indeed they did. When Galileo aimed his telescope at the bright open star cluster we call the Pleiades (M45), he saw points of light nobody in history had ever seen. Until that point, astronomers had classified stars into six brightnesses, called "magnitudes." Galileo became the first to describe a star's brightness as "seventh magnitude."

© Springer International Publishing Switzerland 2016
M.E. Bakich, *Your Guide to the 2017 Total Solar Eclipse*, The Patrick Moore
Practical Astronomy Series, DOI 10.1007/978-3-319-27632-8

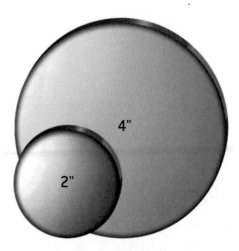

Fig. E.1 A 4-inch diameter mirror has four times the light-gathering area of a 2-inch diameter mirror. Likewise, an 8-inch mirror would collect four times as much light as a 4-inch mirror. *Astronomy* magazine: Roen Kelly

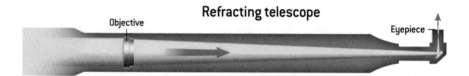

Fig. E.2 A refractor uses lenses—combining at least two, and as many as four, pieces of glass—as their objective (the primary light-gathering device). *Astronomy* magazine: Roen Kelly

2. When I buy my first telescope, will it be complete, or will I have to buy additional items to make it work?

Most telescopes by Celestron and some other manufacturers are complete systems, ready for the sky as soon as you unpack them and perform any recommended setup procedures. But be sure you verify this before you buy. Some models are optical-tube assembly (OTA) only versions. This means what it says: All you're purchasing is the optics and their tube—no mount, tripod, or accessories.

And while the vast majority of OTA instruments on the market are refractors, even some reflectors and catadioptrics now come this way. As a first-time buyer, you may wonder why this is so. The reason stems from the fact that, in many cases, the mount-tripod combination is much more expensive than the optical tube. So, if you're an amateur astronomer who has invested a lot into a mount, you don't need to replace it (or have an extra mount) if you purchase a new telescope. Just swap out the optical tube assemblies on your mount—a task that usually takes only a few minutes—and you're good to go.

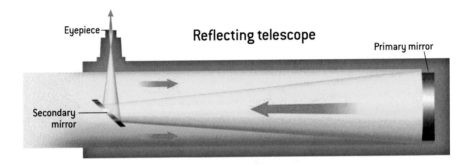

Fig. E.3 A reflector uses mirrors to gather and focus light. In a Newtonian reflector—the most common type—light reflects from the large primary mirror to a smaller secondary mirror. The light's path then bends 90° and enters the eyepiece through a small hole in the tube. *Astronomy* magazine: Roen Kelly

3. I'm interested in observing, but I don't know which scope to buy. What should I do first?

This appendix along with Chapter 14 are a good first steps. Your goal is to learn all you can about telescopes: what types are available, what accessories are the best, and what you'll see through them. Page through this issue and you'll see a range of what's available.

If a telescope interests you, visit manufacturer websites such as www.Celestron. com and read more about it. You'll also find telescope reviews online at *Astronomy* magazine's website, www.Astronomy.com. At first, you may not understand every-thing in the review, but you'll get your first taste of telescope terminology. And you'll learn what's important to veteran observers when they use a telescope. You also will get a feel for optical and mechanical quality, ease of use (including portability), extra features, and—perhaps most importantly—which objects the telescope excels on.

Fig. E.4 A catadioptric, or compound, telescope uses a primary mirror coupled with a correc-tor lens at the front of the tube. The most popular catadioptric telescope is the Schmidt-Cassegrain. *Astronomy* magazine: Roen Kelly

4. Is it true that I should purchase binoculars before a telescope?

No. I used to give this advice to beginning amateur astronomers, but not any more. The view through binoculars—especially from a light-polluted site—often proves disappointing to beginners. However, high-quality binoculars are a valuable observing accessory. Large star clusters look great through binoculars, as do the Milky Way band and the Moon. Oh, and the totally eclipsed Sun, of course!

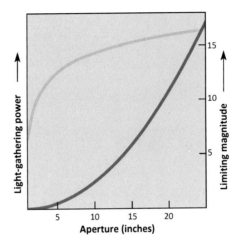

Fig. E.5 A larger scope gathers more light, and therefore, you can observe fainter objects through it. *Astronomy* magazine: Roen Kelly

5. Why are objects I view through my telescope upside-down?

Just before the light collected by your telescope's main lens or mirror enters the eyepiece, it's flipped, so you see an inverted image. A prism assembly (usually called an "image erector") will re-flip the image, but adding this accessory will cause some light to be lost. And regarding observing, the telescope's function is to deliver the maximum amount of an object's light to your eye, so you don't want to lose light from flipping the image. Besides, keep in mind there's no up or down in space, and with most objects, you won't even know they're upside-down. This complaint usually comes from people who use their telescopes to view earthly objects. And if that's what you're doing, an image erector will work great for you because there's more than enough light in the daytime to overcome any light lost through it.

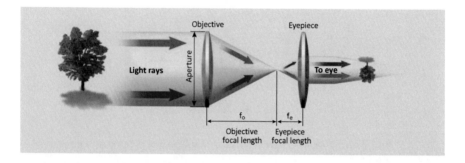

Fig. E.6 Telescope images are usually upside-down, although there is no "up" or "down" in space. *Astronomy* magazine: Roen Kelly

6. Can I use my telescope for views of earthly objects?

Absolutely! Many nighttime observers (usually those with small refractors because of size and portability issues) also use their telescopes for birdwatching or other nature-watching activities. Think about pictures you've seen of sailors in centuries past—they almost always are at the bow looking through a telescope. And here's one more important point: Photographing terrestrial targets is easier in one major way than trying to capture the light from celestial objects. That's because earth-bound objects (we're not talking about animals here) do not appear to move when you view or photograph them through your telescope.

7. Is there any way for me to "test-drive" a telescope?

Yes. Look in your area for an astronomy club, and visit one of its meetings. You'll find others who enjoy astronomy as much as you and are willing to share information and views through their telescopes. At one of the club's public or members-only stargazing sessions, you'll be able to look through many different telescopes in a short period of time.

8. Apart from quality optics, what's the most important thing in a telescope system?

The mount. You can buy the finest optics on the planet, but if you put them on an undersized or poor-quality mount, you won't be happy with your system. No telescope can function in high winds, but a poor mount will transfer vibrations in a light breeze. The mount's quality also affects the "damp-down" time. This is the interval between when you touch the scope (to focus, for example) and when the image in the eyepiece stops moving. Sturdy mounts reduce this to a second or two. Bad mounts increase this time to an intolerable length. Mount quality is one more reason to "try before you buy."

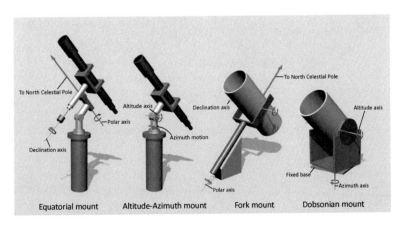

Fig. E.7 An equatorial mount has one axis pointed to the North Celestial Pole. An altitude-azimuth mount moves the telescope horizontal and vertical. A fork mount is similar to an alt-azimuth mount, but the main axis lines up with the North Celestial Pole. A Dobsonian mount is an alt-azimuth type specifically made for Newtonian reflectors. The mount, which combines a cradle and a swivel base, is easy to construct. *Astronomy* magazine: Roen Kelly

9. Is a "go-to" scope better than one without go-to?

Yes. Once properly set up, a go-to scope (actually, it's the telescope's drive that's go-to) will save you lots of time by moving under internal computer control to any object you select. It's important, however, that you understand how to set up your scope, and that will require you to identify a few bright stars. Use a star chart (like the ones you can find at the center of each *Astronomy* magazine) to make this step easier. Even experienced observers prefer go-to scopes because they make star-hopping to deep-sky objects a thing of the past. (When star-hopping, an observer locates a bright star and proceeds to ever-fainter stars until the target appears.)

10. Since telescopes are used outside, do they need electricity?

No, but their motor (go-to) drives do. In most cases, telescope drives use direct current, which means you can either use batteries or an adapter that will allow you to plug into an electrical outlet.

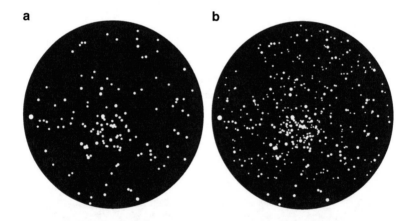

Fig. E.8 (a, b) The faintest star you can see through your telescope—its "limiting magnitude"—depends on the size of the objective. The two illustrations show a star field through a 2-inch scope (*left*) and a 4-inch scope. *Astronomy* magazine: Roen Kelly

11. OK, so what's the best scope for you?

Always remember the first law of telescope buying: The best telescope for you is the one you'll use the most. If it takes an hour to set up a scope each time, or if your scope is large, heavy, and difficult to move, you may observe only a handful of times each year. If, on the other hand, your scope is quick to set up or if you can mount it permanently (or at least its tripod or pier), you may use it several times each week. A small telescope that's used a lot beats a big scope in a closet every time.

Appendix F

30 Cool Facts About the Moon

When you read that title, you probably think, "Sure, I expect this kind of information in a book about a lineup of the Sun, the Moon, and Earth." And so you should. But the real reason for this list was an absolutely astounding television segment that aired January 12, 2015, on the QVC channel. "QVC" stands for "Quality, Value, and Convenience." Fair enough. I guess four letters would have been too many, so they left out "E" for "Education," or "K" for "Knowledge," or "S" for "Science."

Anyway, on that date, viewers were treated to a conversation between host Shawn Killinger and fashion designer Isaac Mizrahi. Killinger picks up a seafoam-colored dress, says that it's a happy color, and then adds, "It almost kinda looks like what the Earth looks like when you're up a zillion miles away from the planet Moon." To her credit, she immediately corrects herself, saying, "From the planet Moon … from *the* Moon," but Mizrahi heard her say it and parrots, "From the planet Moon."

© Springer International Publishing Switzerland 2016
M.E. Bakich, *Your Guide to the 2017 Total Solar Eclipse*, The Patrick Moore
Practical Astronomy Series, DOI 10.1007/978-3-319-27632-8

Fig. F.1 The Moon is our planet's only natural satellite. NASA

That's when the discussion goes from "Oops" to worse. "Isn't the Moon a star?" Killinger asks. "No, the Moon is a planet, darling," Mizrahi replies. After that, it's back and forth with the only notable addition being Killinger asking, "Isn't the Sun a star?" and Mizrahi replying, "I don't know what the Sun is. We don't know what the Sun is."

You get the idea. The clip went viral. It aired all over the Internet, on major news programs, and, of course, on late-night comedy shows. Well, all that publicity gigged me into producing this list.

To be fair, not everyone remembers details they may have last heard in third grade. But if you want to understand everything about the 2017 eclipse—and especially if you want to tell others about it—at least a small working knowledge about the objects involved is necessary. To that end, here are 30 brief facts about the Moon in no particular order (well, except for the first one).

#1. The Moon is not a star or a planet, it's a satellite. In fact, it's Earth's only natural satellite. (We distinguish the Moon from the vehicles we launch into space by calling them artificial satellites.) So, planets (like Earth) orbit stars (like the Sun), and satellites orbit planets. Because we know the [capital-M] Moon so well, often you'll hear astronomers call satellites that orbit other planets "moons," but they spell that word with a lower-case "m."

#2. The Moon came from Earth. It didn't form beside our planet, and Earth's gravity didn't capture it. The latest theory says that the Moon formed some 4½ billion years ago from debris blasted out of Earth by the impact of a body the size of Mars.

#3. The Moon and the Sun appear to be the same size. This works because the Sun's diameter is 400 times greater than the Moon's, but the Sun also lies 400 times as far away.

#4. Here's a two for the price of one: The Moon takes 27.3 days to orbit Earth once, but 29.5 days to go from New Moon back to New Moon. The reason these two periods are different is because, at the same time the Moon is orbiting Earth, Earth is orbiting the Sun. So, it takes an extra 2.2 days for the three objects to once again line up at New Moon (or, really, from any phase you choose to the next time that phase appears).

#5. The Moon's surface reflects only 12 percent of the light that falls on it. And you thought the Full Moon was bright!

#6. The Moon always keeps the same face toward Earth. Because it does, it must rotate (that is, spin) once for each orbit it makes. This may be the single biggest misconception about the Moon. Many people think that because the same face always points at us, the Moon doesn't rotate. That's false, and it's easy to show how this works. Put a chair in the middle of a room. That chair will represent Earth, and you will be the Moon. Stand a few feet away from it, and face it. Also notice which wall in the room you're facing. Next, orbit halfway around the chair, always keeping your front side toward it, and stop. A quick glance at the room will show that you're now facing the opposite wall. Aha! For this to happen you must have rotated one-half spin. Continue around to your starting point, and you'll see that as you orbited once, you also rotated once. Pretty cool, eh?

#7. The Moon's average distance from Earth is 238,000 miles (383,000 kilometers). It can be as far away as 252,724 miles (that's 406,720 kilometers) or as near as 221,439 miles (356,372 kilometers). So, the Moon can show a 14 percent change in size between the two extremes. At its average distance, and at a person's average pace, it would take 9 years to walk to the Moon.

#8. The Moon's diameter is 2,159 miles, or 3,474 kilometers. Only four satellites in our solar system are larger: Ganymede, Titan, Callisto, and Io.

#9. Here's one related to eclipses: The Moon's orbit tilts 5.14° to Earth's orbit around the Sun. This is the reason we don't have solar eclipses at every New and lunar eclipses at each Full Moon.

#10. Of the 9,113 official features on the Moon, a mere 421 are not craters. That's only 4.6 percent.

#11. The Moon's temperature can vary from 242° Fahrenheit in a sunny spot to −334° Fahrenheit in a shady spot near one of its poles. On the Celsius scale, that's a range from 117° to −203°.

#12. The ratio of the sizes of Earth and the Moon—27.6 percent—is far larger than any other planet/satellite combination. The next nearest ratio—5.5 percent—describes Neptune and its largest moon, Triton.

#13. The Moon is currently moving away from Earth at a rate of about 1½ inches per year, with an accuracy of about 1/32 of an inch.

#14. The Moon is the major influence for ocean tides on Earth. Our satellite has twice as much effect on the tides as the Sun does.

#15. The Full Moon, though bright, is only 1/400,000 as bright as the Sun. If the entire sky were covered with Full Moons, we would receive only about one-fifth the illumination of the Sun on a bright day.

#16. It wasn't Columbus who proved Earth was round. In the fourth century B.C., Greek philosopher Aristotle showed that Earth was a sphere by noting that our planet's shadow always appears round during lunar eclipses.

#17. All parts of the Moon see the same amount of sunlight and darkness. So, although I am a Pink Floyd fan, I'm sorry to say there is no dark side of the Moon. 50 percent of it is always dark, but the areas covered continually change as sunrise and sunset occurs.

#18. I find this one fascinating. Although the First and Last Quarter Moons show us 50 percent of the lit area of a Full Moon, each is only about 10 percent as bright as the Full Moon. The reason? The angle at which sunlight strikes the Moon.

#19. Earthshine, the "lighting up" of the Moon's dark area a few days before or after New Moon, happens because sunlight reflects off Earth and falls on the Moon, slightly illuminating the part we normally wouldn't see. Earthshine also goes by the names "ashen glow" and "the old Moon in the new Moon's arms."

#20. Astronomers call the dividing line between the light and dark parts of the Moon the terminator. Between New Moon and Full Moon (that is, as the Moon's face is waxing, a word that means "growing"), the terminator shows where sunrise is happening. Between Full Moon and New Moon (when the Moon's face is waning, a word that means "shrinking"), it is the line of sunset.

#21. A total of 12 American astronauts have walked on the Moon. They traveled there as part of missions Apollo 11 through Apollo 17. Apollo 13 had technical problems and never landed on the Moon. The first landing occurred July 20, 1969. Neil Armstrong became the first man on the Moon the next day. The sixth and final landing occurred December 11, 1972. Eugene Cernan became the last man on the Moon when he stepped off its surface December 14.

#22. The Moon is just as much a daytime object as a nighttime one. You can see it in the daytime half of each lunar month. And here's a bonus fact: on average, the Moon rises 50 minutes later each day.

#23. The Moon does contain some water. In 2009, the Lunar Crater Observation and Sensing Satellite (LCROSS, for short) found water ice in Cabeus Crater, which lies only 60 miles from the Moon's South Pole.

#24. The surface gravity of the Moon is 16.6 percent that of Earth's surface gravity. Said another way, if you weigh 100 pounds on Earth, you would weigh 16.6 pounds on the Moon—not counting your spacesuit, helmet, boots, air tanks, and everything else you would need to survive.

#25. The six successful Apollo missions returned nearly 838 pounds of lunar rock and soil in 2,196 separate samples.

#26. The so-called "Super Moon" is just the nearest Full Moon during any calendar year. In 2015, the Super Moon occured Sunday, September 27. Wait, that was the same date as a total lunar eclipse. Awww, that's gotta mean something, right?

#27. Because of tidal forces between Earth and the Moon, our day is getting longer by 1.4 milliseconds each century.

#28. The Moon's atmosphere is less than one hundred-trillionth the density of Earth's atmosphere at sea level.

#29. The lighter regions of the Moon are highlands. The darker regions, often called "seas" or "maria," are basins filled with dark lava.

#30. According to a 2011 study by astronomers, 8.3 percent of Earth-like planets are likely to have a Moon-like satellite.

Appendix G

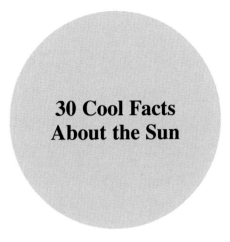

30 Cool Facts
About the Sun

An eclipse also involves the Sun, so here's the bookend to the previous appendix. I hope you enjoy (and slightly marvel at) these 30 brief facts about the Sun in no particular order.

© Springer International Publishing Switzerland 2016
M.E. Bakich, *Your Guide to the 2017 Total Solar Eclipse*, The Patrick Moore
Practical Astronomy Series, DOI 10.1007/978-3-319-27632-8

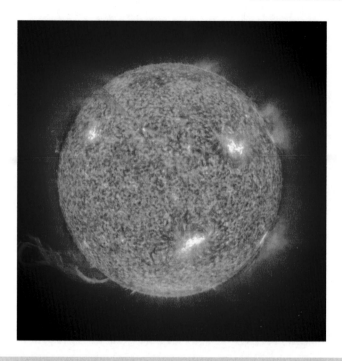

Fig. G.1 The Sun is the center of our solar system and the source of life on Earth. The Solar and Heliospheric Observatory, launched into low-Earth orbit in December 1995, took this image through a Hydrogen-alpha filter. NASA/ESA/SOHO

1. The Sun is a star. It's similar to all the stars we see at night. Of course, some of them are hotter, and some are cooler. Some are bigger, some smaller. Some have more mass than the Sun, some less. But the biggest difference between our Sun and all those stars is distance. Even the next-closest star—Alpha Centauri—is more than 250,000 times as distant than the Sun. And our North Star? It's more than 20 million times as far away!
2. The Sun is so large that 1.3 million Earths could fit inside it. And if you want to compare it to Jupiter, some 1,300 giant planets of that size could fit within our daytime star.
3. The Sun's diameter is about 865,400 miles, or 1,392,700 kilometers. That makes it 109 times Earth's diameter.
4. The Sun's mass—that is, the amount of stuff that makes it up—is 330,000 times the mass of Earth. In fact, the Sun accounts for 99.86 percent of the mass of our entire solar system. That's why we call it a solar system, from the Latin word "sol," meaning Sun.
5. Astronomers estimate that the Sun formed from a huge condensing cloud of dust and gas 4.567 billion years ago.
6. The Sun's angular size, or how big it appears to us on Earth, ranges from 31.6 arcseconds to 32.7 arcseconds. The average of those extremes equals the size of a quarter 8.85 feet, or 2.7 meters, away.

7. The Sun's magnitude, or apparent brightness, is −26.74. In common terms, this is some 13 billion times brighter than the next-brightest star, Sirius, in the constellation Canis Major the Great Dog. Let me expand on this a bit. We think of the Full Moon as bright, but its light really is nothing compared to the Sun. The Sun is 1 million times as bright as the Full Moon. In fact, if our sky were filled with Full Moons, they would provide only 20 percent of the Sun's light.

8. Speaking of sunlight, it's composed of approximately 40 percent light, 50 percent heat, and 10 percent ultraviolet radiation. Well, in space, anyway. By the time sunlight reaches Earth's surface, our atmosphere has filtered out some 70 percent of the ultraviolet light.

9. Most people think that the Sun is yellow, but it is, in fact, a white star. It is obviously white from space or when it rides high in our sky. When it approaches the horizon, however, our atmosphere bends the shorter wavelengths (that is, the colors green, blue, and violet) away from our eyes, making a low Sun appear yellow, orange, or red.

10. The Sun's average surface temperature is 9,941° Fahrenheit, or 5,504° Celsius. The core's temperature, on the other hand, is 28.3 million degrees Fahrenheit, or 15.7 million degrees Celsius.

11. Astronomers call the Sun's visible surface the photosphere. Its thickness ranges from tens of miles to a few hundred miles.

12. The chromosphere is a layer about 1,200 miles thick that lies above the photosphere. "Chromo" means "color," and this region got its name from the reddish flash often seen just before and just after totality during a total solar eclipse.

13. Above the chromosphere is the corona. It has an average temperature between 1.8 and 3.6 million degrees Fahrenheit, but its hottest regions zoom to 36 million degrees Fahrenheit.

14. The Sun emits a continuous stream of plasma called the solar wind. On Earth, the most visible effect of the solar wind is the production of the northern and southern lights, the aurorae. Intense bursts of solar wind also can disrupt communications, especially those that route through orbiting satellites.

15. Sunspots appear dark because can be as much as 2,700° Fahrenheit cooler than the surrounding surface. They are concentrations of the Sun's magnetic field where it breaks through the photosphere causing the region to radiate less energy.

16. Records of sunspots date to the 4th century B.C. During the past 100 years, observers have reported between 40,000 and 50,000 sunspots. The numbers ebb and flow with a fairly steady 11-year period. Solar scientists dub each 11 years a "solar cycle." The first peaked in 1760. Since then, 23 cycles have come and gone. Solar cycle 24 began in June 2009.

17. The Maunder minimum is a 70-year period, from 1645 to 1715, when sunspots all but disappeared during a diminished period of solar activity.

18. One of the best-known of all astronomical facts is that the Sun lies an average of 93 million miles from Earth. Astronomers call this average distance an astronomical unit. In 2012, the International Astronomical Union defined the length of the astronomical unit as 149,597,870.7 kilometers, or 92,955,807.3 miles. But Earth's orbit isn't a perfect circle. We're closest to the Sun around January

4 each year and farthest from it around the Fourth of July, with the difference of the two extremes being about 3 million miles, or 5 million kilometers.

19. The Sun (and the rest of the solar system) lie 27,200 light-years from the center of our galaxy, the Milky Way. It takes our solar system between 225 million and 250 million years to complete just one orbit within the Milky Way. And that's traveling at the incredible speed of 135 miles per second. Speaking of the Milky Way, it contains a total of between 250 billion and 400 billion stars.

20. The Sun is almost a perfect sphere. Its polar diameter is only 6 miles less than its equatorial diameter.

21. The Sun does rotate, but different parts of it spin at different speeds. It takes 25.6 days for the equatorial regions to rotate once, and 33.5 days for the polar regions to do so.

22. The Sun is mainly hydrogen and helium. Hydrogen makes up 73.46 percent of the Sun's mass, and helium accounts for 24.85 percent. That leaves just 1.59 percent for everything else. The next three elements by mass are oxygen (0.63 percent), carbon (0.22 percent), and neon (0.17 percent).

23. Because Earth rotates from west to east, the Sun appears to rise in the east in the morning and set in the west at night. By the way, this doesn't reverse in the Southern Hemisphere. The Sun still rises in the east and sets in the west there.

24. The Sun's surface gravity is 28 times that of Earth's.

25. Each 11 years, the Sun's magnetic polarity reverses. This ties in directly with the 11-year sunspot cycle.

26. Astronomers have figured out the Sun's movement through our galaxy and you can see the point toward which we're moving. If you go out on a clear night in the Northern Hemisphere's summer, look toward the magnitude 4.4 star Nu Herculis, which lies near 18 hours right ascension and 30' north declination. This star lies approximately 12° southwest of 1st-magnitude Vega in the constellation Lyra the Harp.

27. The first solar flare ever observed, on September 1, 1859, by British astronomer Richard Carrington, was also the most powerful.

28. From Earth, the Sun appears 30 times larger and 900 times brighter than from Neptune.

29. The outer corona of the Sun extends 12 times our star's radius from its surface.

30. The Sun travels along an imaginary line in our sky called the ecliptic, because that's the only place eclipses can occur. The group of constellations surrounding the ecliptic is known as the zodiac.

Appendix H

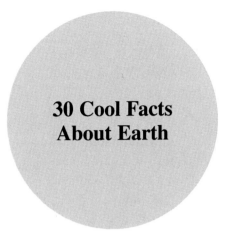

30 Cool Facts
About Earth

This appendix deals with the third body necessary for a total solar eclipse to occur: Earth. And just like the other two, it presents tidbits of knowledge about the third rock from the Sun—just in case someone asks.

© Springer International Publishing Switzerland 2016

M.E. Bakich, *Your Guide to the 2017 Total Solar Eclipse*, The Patrick Moore
Practical Astronomy Series, DOI 10.1007/978-3-319-27632-8

Fig. H.1 Earth is home to humanity and the only location in the universe where we know life exists. NASA

1. Earth is "Earth." It's not "the" Earth. It's just Earth. That's its name. You wouldn't say "the Mars" when you're talking about the Red Planet. It's just Mars. Lots of people and lots of publications get this wrong. Is it a big deal? No, but now you know.
2. Earth revolves around (orbits) the Sun, once a year, at a speed of 66,616 mph. That speed would get you to the Moon in about 3½ hours.
3. Earth rotates on its axis once every 23 hours 56 minutes and 4 seconds. Because of its rotation, Earth's equatorial diameter is 27 miles greater than its polar diameter.
4. Traces of Earth's atmosphere exist as far as 400 miles from the surface.
5. 71 percent of Earth's surface is covered by water. 97 percent of that water is salt water in the oceans, with 1 percent left as fresh water in lakes, rivers, and the like. The other 2 percent are in glaciers, free ice, and underground. Even with all that water, it makes up only 7 one-hundredths of one percent of Earth by mass.
6. Earth's core has a temperature of approximately 10,800° Fahrenheit, about the same as the temperature of the Sun's surface. That's amazing!
7. Earth's day is getting longer by 1.7 milliseconds each century.

8. The Amazon rainforest produces 20 percent of Earth's atmospheric oxygen. And oxygen is what makes our world unusual among the planetary family. We haven't found nearly as much free oxygen on any other planet. By volume, our air is 21 percent oxygen. For contrast, the atmospheres of Venus and Mars contain merely a trace of this element.

9. Earth formed into a primordial planet (looking nothing like it does now) approximately 4.54 billion years ago.

10. Earth has a well-established magnetic field that originates in its core. Our planet's core has two parts: an inner solid region and an outer liquid one. Both are made of a nickel-iron alloy, but only the inner core is under enough pressure to remain solid. Flow currents in the outer core, caused by heating from the inner core and by Earth's rotation, create a dynamo that generates the magnetic field. Generally, our magnetic field extends between and 30,000 and 40,000 miles above Earth's surface in the direction toward the Sun and between 200,000 and 225,000 miles in the direction opposite the Sun. The difference is due to the pressure exerted on the magnetosphere by the solar wind.

11. Earth's rotation causes everything in the sky to appear to move from east to west at a rate of 15° per hour.

12. In 2012, the International Astronomical Union defined the average distance between Earth and the Sun—a length called the astronomical unit—as exactly 149,597,870,700 meters. That works out to 92,955,807 miles. So, for those of us who, from childhood, have been saying, "The average distance between Earth and the Sun is 93 million miles," we've been accurate to more than one-twentieth of one percent!

13. Currently, Earth's axis tilts at an angle of 23.44°. In the past 5 million years, that angle has varied from 22° to 24.5° approximately every 41,000 years. The planet with a tilt closest to ours is Mars, which tilts 25.19°. The planet with the greatest tilt is Venus, with a whopping 177.36° tilt.

14. Of all the planets in our solar system, Earth has the greatest density, at 5.51 grams per cubic centimeter. This means that, on average, Earth weighs 5.51 times as much as an equal volume of water.

15. Every day, approximately 100 tons of cosmic dust (and also a few larger pieces we call meteors) enters our atmosphere, eventually falling to the surface.

16. The highest recorded air temperature on Earth—134.1° Fahrenheit—occurred July 10, 1913, at Furnace Creek Ranch in Death Valley, California. The lowest recorded air temperature—−128.6° Fahrenheit—was measured July 21, 1983, at Vostok Station in Antarctica.

17. Although Earth is *almost* a sphere, "almost" only counts in horseshoes and hand grenades. Scientists define our planet's shape as an oblate spheroid, which is a consequence of its rotation. Because of that little bit of stretching at its waist, Earth's equatorial diameter is 26.6 miles greater than its polar diameter.

18. If all of Earth's ice melted, the sea level would rise 330 feet.

19. Earth's mass is 6.58 sextillion tons. That's the number 658 followed by 19 zeroes. If you add up the masses of Mercury, Venus, and Mars, they equal only 97.7 percent of the mass of our planet.

20. Earth's albedo, the amount of light it reflects, is 37 percent, sixth-best out of the planets in our solar system. The most reflective planet is Venus, with an albedo of 65 percent. The least reflective world is rocky Mercury. It reflects only 10 percent of the light that strikes it.

21. The eccentricity of Earth's orbit, that is, how much its orbit varies from a circle, is 1.67 percent, third-best among the planets. Venus' orbit is closest to a circle. It varies by just 0.68 percent. Mercury and Pluto have the most eccentric orbits. Mercury's is 20.6 percent off circular, and Pluto's is a whopping 24.8 percent from perfect.

22. For a manmade object to escape Earth's gravity, it has to be traveling at a speed of 25,009 mph. This speed is called the escape velocity.

23. Earth spins pretty quickly. A person at the equator is traveling at a speed of 1,038 mph. For those living in Cairo, Egypt, or New Orleans, that is, at a latitude of 30°, the speed would be 899 mph. At 40° latitude—which could be Beijing, China, or THE Ohio State University in Columbus, Ohio—it would be 795 mph. And Londoners, at a latitude of 51.5°, are moving at 646 mph.

24. Earth has a volume of some 260 billion cubic miles. That means six and two-thirds Mars-sized bodies could fit inside it. And it would take nearly 18 Mercurys to fill it up. Still, as big as our world is, 1,266 Earths could fit inside Jupiter.

25. You may have heard that the Great Wall of China is the only manmade object visible from space. It is not. China's air pollution, however, is.

26. Pilots or astronauts must wear pressurized suits at altitudes of 12 miles or higher. That's because at that height water boils at body temperature. This altitude is called the Armstrong limit. But it's not named for the Moon-walking astronaut. Rather, the name comes from the first person to recognize that this would happen to a person at that altitude, Harry George Armstrong. He founded the U.S. Air Force's Department of Space Medicine in 1947.

27. If you formed a sphere out of all the water on Earth, it would have a diameter of 534 miles.

28. Roughly 99 percent of all of Earth's gold lies within the core. If you could extract it, that much gold would form a layer covering the Earth 1½ feet deep.

29. Earth's crust is composed of seven or eight major tectonic plates and many minor ones. The relative movement between these plates varies from zero to 4 inches per year.

30. Earth, Texas, is the only place on Earth named Earth.

Appendix I

Take a Break with an Eclipse Word-Finder Puzzle

Some things related to the upcoming eclipse that you can do, either alone or with family or friends, should be fun. This word finder puzzle is one example. Find these 45 words or word combinations:

Altitude	Disk	Partial	StJoseph
Aphelion	Eclipse	Penumbra	Sun
Apogee	Ecliptic	Perigee	Sunspot
Azimuth	Filter	Perihelion	Syzygy
BailysBeads	FirstContact	Photosphere	Telescope
Binoculars	Flare	Prominence	ThirdContact
Centerline	FourthContact	Rosecrans	Total
Chromosphere	Magnitude	Saros	Totality
Corona	Moon	SecondContact	Umbra
Crescent	NewMoon	Shadow	
Darkness	Node	ShadowBands	
DiamondRing	Obscuration	Solar	

© Springer International Publishing Switzerland 2016
M.E. Bakich, *Your Guide to the 2017 Total Solar Eclipse*, The Patrick Moore
Practical Astronomy Series, DOI 10.1007/978-3-319-27632-8

```
Q D G C A L X Z R Z J K Q E T E Z S X Y L F H P L Z V Q G D V S D
S S W Y C C L E V C X A L P S Q S B I N O C U L A R S P R X L M A
O R W N E P A K V G F I R S T C O N T A C T J J J Z G E A C T Q J A
L Z P F N W Y N U M H Z C B N L R D Y A P X P Q R S N M W Y S S T
A P E S T C X Z W P T F R S M M D L W C Q L P Y P O P K F Y H R H
R U K N E T C S U N S P O T H U Q U E T A T H I R D C O N T A C T
D X A A R L T V M W R F D U C A U R C V F Q L O B G P Y O N D P X
E E J R L O U S F I E W N U R R D K C H W C C J V G R Z E L O V Q
V T C C I D P I H Q U O X D W T Y O P T E Q D B Q U O W Z N W V U
W Z O E N C L E G V I O A N O C H W W P C L O M Y W M P H K K M M
E H S S E T A F L L E U S M E N C C G B A A Z U V O I F P X U P Q
B C G O E X H O N O I T A R U C S B O G A R A S O T N I W F T D E
F I E R E H P S O T O H P T Q I P K O N N N T N L Y E D A H O T E
X E E I F Z I O Y O S M U D O K Q D M X T I D I R O N X J E Q J G
Y S N R H P Y K E C L Z O X S Q I K P O F A R S A U C S L Z N E I
N S D A E B S Y L I A B D H T E C B N Y W N C D I L E E U X P X R
E D E V I Q O F U B R C S T J O S E P H Q P G T N W S K D E W J E
N P S Y B P G B T L X H J Y X X G W Y N J K Y C B O T L N O W N P
V O T Q F X M G D A E R T A Z F S A L S Q L R M R P M U S H N E U
K K O E Z A G A H T T O C L I Y D S C M E V O V J E M A M I J W A
H E I M L C R T H O I M A Q L W G B T Y A L K Z P B S F I N Y P Y
L C E S A E Y L A T F O T G U B R Y K Z T L J J R N L C X D H G N
P L X S Q C S Q N V A S N Q R C J I L R X K D A N A R K E E A G O
Y I Y E M Y V C T F T P O Q M N Q G K Q R U N N R W W S L N L N R
V P S N D A K Q O K H H C R I L D F J P X C W E H A V I E X T X B
P T P K A X G G U P M E D J U P E J P H M G D X Y C O Y Y Q I D Y
V I A R M M I N T R E R N N O I L E H I R E P D M N E D F V T L V
F C Z A W J P E I X Z E O T C S A Y T I L A T O T E V I K L U L A
X T I D I L O A B T D P C I K S L E H U N G N P G F N I A V D H J
E I M C O J G F H S U H E W Q Y B F K P L U S O K S O R A S E L Z
D J U Z M Y A G H Z W D S L H A A R O Z S P P Z Z L O Q M W Q H S
R V T I A Z B D H Y L T E Y R N H C U R R A H Z K F W M D I S K U
D A H N Z E K K Y Y D B B U B Z G Z C K X Z Z X V B X Q B M Y T A
```

Fig. I.1

Appendix J

A Collection of Postage Stamps Featuring Past Eclipses

Total solar eclipses are big events. They generate interest among the public, science courtesy of researchers, and cash thanks to tourism. During the past 50 years, postal services around the world have realized they could add to their coffers by producing souvenirs for collectors. And philatelists have responded.

I'm one of them. All of the stamps in this appendix are from my personal collection. I acquired some of them in distant lands while leading solar eclipse expeditions. Others, quite honestly, came from eBay. It's a pretty complete collection. Here are some of the highlights. These aren't necessarily the rarest, just the prettiest in my opinion.

© Springer International Publishing Switzerland 2016
M.E. Bakich, *Your Guide to the 2017 Total Solar Eclipse*, The Patrick Moore
Practical Astronomy Series, DOI 10.1007/978-3-319-27632-8

Fig. J.1 Romania: February 15, 1961

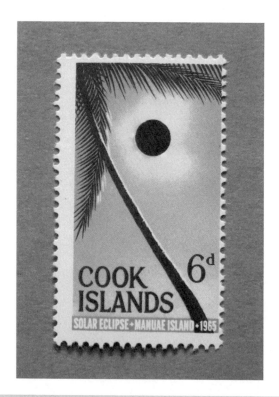

Fig. J.2 Cook Islands: May 30, 1965

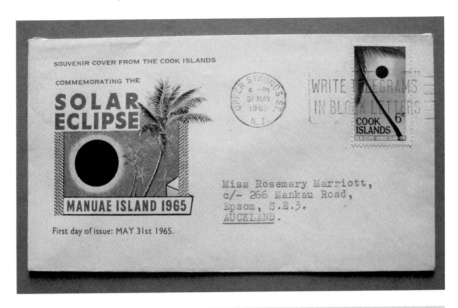

Fig. J.3 Manuae Island: May 30, 1965

Fig. J.4 Mexico: March 7, 1970

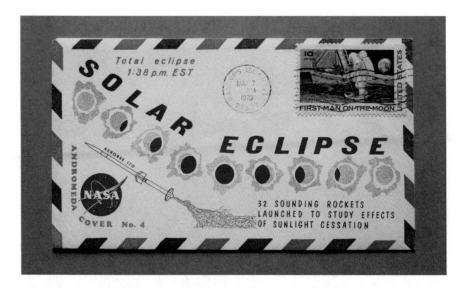

Fig. J.5 United States: March 7, 1970

Fig. J.6 Niger: June 30, 1973

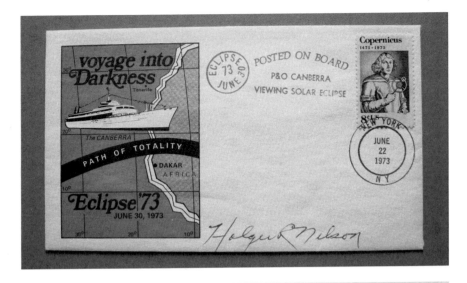

Fig. J.7 Aboard the Canberra: June 30, 1973

Fig. J.8 Senegal: June 30, 1973

Fig. J.9 Aboard the Fairwind: October 12, 1977

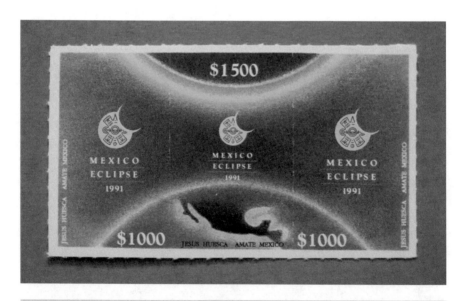

Fig. J.10 Mexico: July 11, 1991

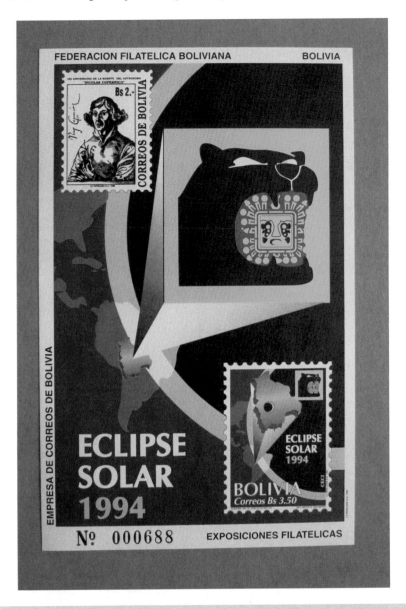

Fig. J.11 Bolivia: November 3, 1994

Fig. J.12 Thailand: October 24, 1995

Fig. J.13 Viet Nam: October 24, 1995

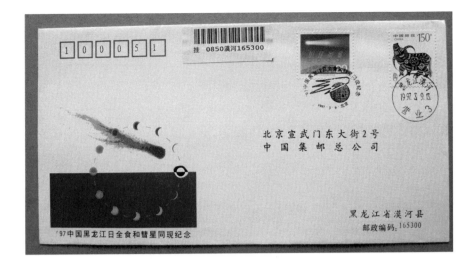

Fig. J.14 China: March 9, 1997

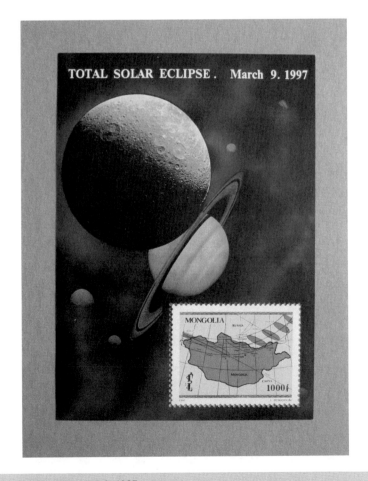

Fig. J.15 Mongolia: March 9, 1997

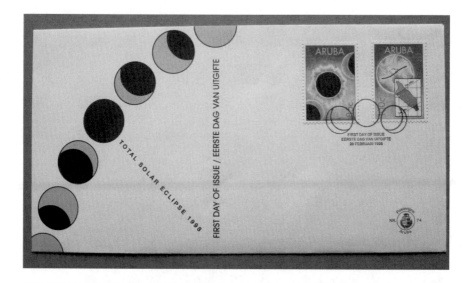

Fig. J.16 Aruba: February 26, 1998

Fig. J.17 Guinea: February 26, 1998

Fig. J.18 Montserrat: February 26, 1998

Fig. J.19 Netherlands Antilles: February 26, 1998

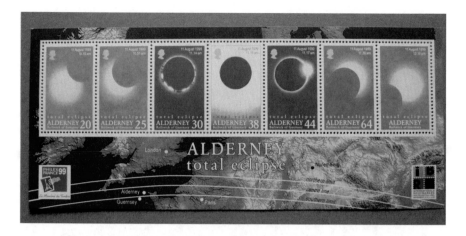

Fig. J.20 Alderney: August 11, 1999

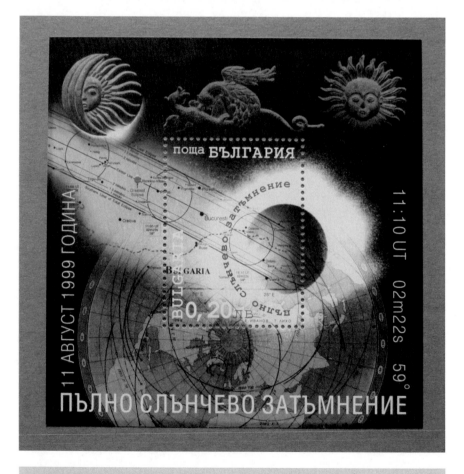

Fig. J.21 Bulgaria: August 11, 1999

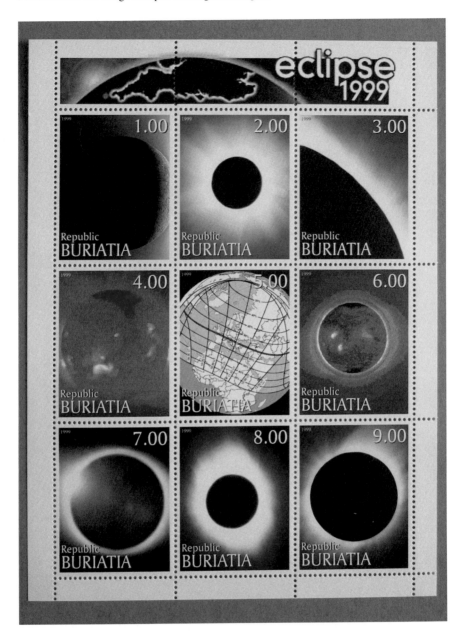

Fig. J.22 Buriatia: August 11, 1999

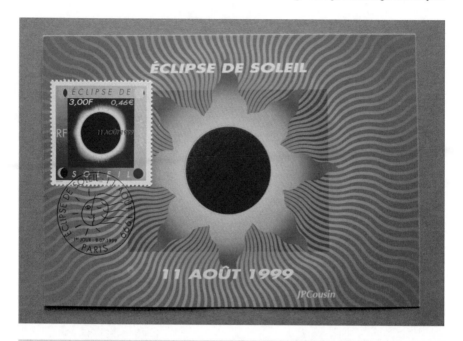

Fig. J.23 France: August 11, 1999

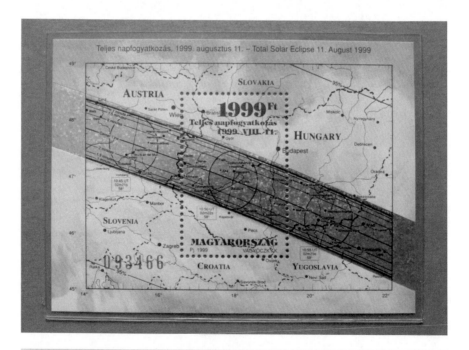

Fig. J.24 Hungary: August 11, 1999

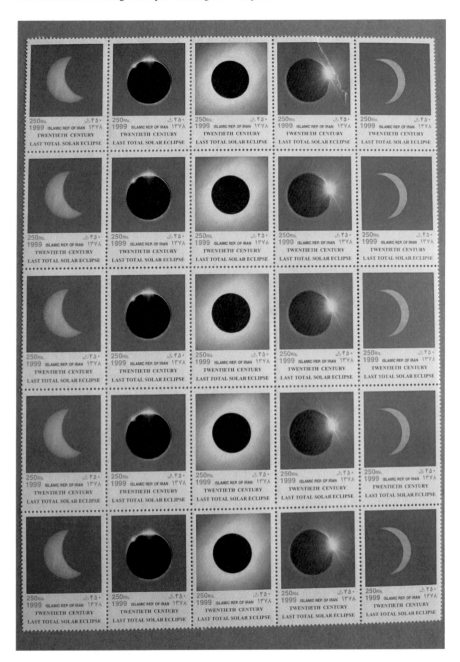

Fig. J.25 Iran: August 11, 1999

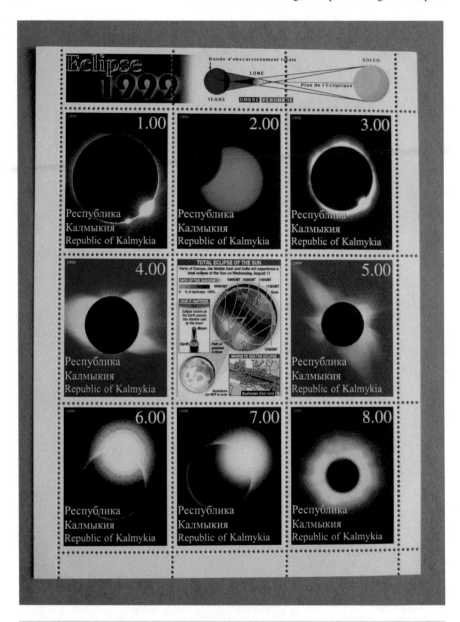

Fig. J.26 Kalmykia: August 11, 1999

Fig. J.27 Maldives: August 11, 1999

Fig. J.28 Romania: August 11, 1999

Fig. J.29 Zimbabwe: June 21, 2001

Fig. J.30 Angola: December 4, 2002

Fig. J.31 South Africa: December 4, 2002

Fig. J.32 Pitcairn Island: April 8, 2005

Fig. J.33 Egypt: March 29, 2006

Fig. J.34 Libya: March 29, 2006

Fig. J.35 Turkey: March 29, 2006

Fig. J.36 China: July 22, 2009

Appendix K

A Complete State-by-State List of Locations on the Centerline

In this appendix, I have listed every location (city, town, village, burg, and point of interest) within about 5 miles of the centerline. In other words, these are the prime locations. Rather than alphabetically, I ordered this list according to which location the Moon's shadow covers first.

Oregon

All locations in Oregon have durations of totality between 1m59s and 2m10s.

Depoe Bay
Lincoln Beach
Salishan Spa & Golf Resort
Falls City
Monmouth
Independence
Ankeny National Wildlife Refuge
Turner
Aumsville
Sublimity
Stayton
Mehama
North Santiam State Park
Mill City
Gates

© Springer International Publishing Switzerland 2016
M.E. Bakich, *Your Guide to the 2017 Total Solar Eclipse*, The Patrick Moore
Practical Astronomy Series, DOI 10.1007/978-3-319-27632-8

Detroit Lake State Park
Detroit
Warm Springs
Madras
John Day Fossil Beds National Monument
Prairie City
Dixie Butte
Bates
Unity Reservoir
Huntington

Idaho

All locations in Idaho have durations of totality between 2m10s and 2m19s.

Crane Creek Reservoir
Lightning Ridge
Grandjean Campground
Stanley
Redfish Lake Lodge
Mackay Reservoir
Mackay
Howe
Terreton
Mud Lake
Market Lake Wildlife Management Area
Roberts
Menan
Rexburg
BYU—Idaho
Kelly Canyon Ski Resort
Driggs-Reed Memorial Airport
Driggs
Victor

Wyoming

All locations in Wyoming have durations of totality between 2m19s and 2m30s.

Jackson Hole Mountain Resort
Jackson Hole Airport
Moose

Ross Lake
Casper-Natrona County International Airport
Vista West
Mountain View
Hartrandt
Mills
Casper
Evansville
Casper Mountain
Ayres Natural Bridge Park
Glendo
Glendo Reservoir
Glendo State Park

Nebraska

All locations in Nebraska have durations of totality between 2m30s and 2m39s.

Alliance
Tryon
Arnold
Callaway
Pressey State Wildlife Management Area
Pleasanton
Ravenna
Cairo
Aida
Grand Island
Doniphan
Phillips
Henderson
Sutton
Morphy Lagoon National Wildlife Management Area
Fairmont
Geneva
Exeter
Wilber
Beatrice
Kirkmans Cove Recreation Area
Falls City

Kansas

Every location in Kansas has a duration of totality in the 2m39s range.

Hiawatha
Highland
Troy
Wathena
Elwood

Missouri

All locations in Missouri have durations of totality between 2m39s and 2m41s.

Brown Conservation Area
Wolf Creek Bend
Riverbreaks Conservation Area
Monkey Mountain Conservation Area
Rosecrans Memorial Airport
St. Joseph
Country Club
Easton
Gower
Smithville Lake State Wildlife Area
Plattsburg
Lathrop
Holt
Lawson
Crooked River Conservation Area
Hardin
Norborne
Carrollton
Waverly
Grand Pass Conservation Area
Van Meter State Park
Marshall
Arrow Rock State Historic Site
Blackwater
New Franklin
Boonville
Davisdale Conservation Area
Rocheport
Big Muddy National Fish and Wildlife Refuge
Overton Bottoms Conservation Area

Eagle Bluffs Conservation Area
Plowboy Bend Conservation Area
Columbia
Rock Bridge Memorial State Park
Three Creeks Conservation Area
Ashland
Columbia Regional Airport
New Bloomfield
Holts Summit
Chamois
Portland
Ben Branch Lake Conservation Area
Lost Valley Lake Resort
Long Ridge Conservation Area
Union
St. Clair
Lonedell
Richwoods
Hillsboro
De Soto
Hematite
Olympian Village
Festus
Valles Mines
Magnolia Hollow Conservation Area
Ste. Genevieve
St. Mary
Perryville
Seventy-Six Conservation Area
Frohna
Altenburg

Illinois

Every location in Illinois has a duration of totality in the 2m41s range.

Prairie Du Rocher
Chester
Middle Mississippi River National Wildlife Refuge
Lake Murphysboro
Lake Murphysboro State Park
Murphysboro
Harrison
Pomona

Carbondale
Evergreen Park
Makanda
Cobden
Giant City State Park
Crab Orchard National Wildlife Refuge
Goreville
Ferne Clyffe
Vienna
Camp Ondessonk
Golconda Marina

Kentucky

Every location in Kentucky has a duration of totality in the 2m41s range.

Hampton
Burna
Smithland
Tiline
Kuttawa
Eddyville
Mineral Mounds State Park
Princeton
Jones-Keeney Wildlife Management Area
Pennyrile Forest State Resort Park
Cerulean
Hopkinsville
Hopkinsville-Christian County Airport
Jefferson Davis State Historic Site
Pembroke
Trenton
Elkton
Guthrie
Olmstead
Adairville

Tennessee

All locations in Tennessee have durations of totality between 2m41s and 2m39s.

Adams
Cedar Hill
Springfield

Orlinda
Cross Plains
Greenbrier
Ridgetop
White House
New Deal
Portland
Cottontown
Gallatin
Boxwell Scout Reservation
Bethpage
Castalian Springs
Hartsville
Lebanon
Carthage
Defeated Creek Park Campground
Brush Creek
Gordonsville
Chestnut Mound
Granville
Buffalo Valley
Edgar Evins State Park
Baxter
Burgess Falls State Park
Sparta
Doyle
Virgin Falls State Natural Area
Pleasant Hill
Lake Tansi
Cumberland Mountain State Park
Grandview
Spring City
Evensville
Watts Bar Nuclear Power Plant
Ten Mile
Decatur
Athens Regional Park
Niota
Sweetwater
Athens
Englewood
Madisonville
Tellico Plains
Bald River Falls

North Carolina

Every location in North Carolina has a duration of totality in the 2m39s range.

Huckleberry Knob
Joyce Kilmer Memorial Forest
Robbinsville
Marble
Andrews
Topton
Hayesville
Standing Indian Campground
Otto
Highlands
Georgia
Rabun Gap

Georgia

Every location in Georgia has a duration of totality between 2m38.9s and 2m38.7s.

Dillard
Mountain City
Black Rock Mountain State Park
Clayton
Tiger

South Carolina

Every location in South Carolina has a duration of totality between 2m39s and 2m34s.

Mountain Rest
Oconee State Park
Walhalla
West Union
Keowee Key
Seneca
Utica
Six Mile
Clemson

Central
Liberty
Pendleton
Northlake
Centerville
Anderson
Piedmont
Williamston
Belton
Honea Path
Donalds
Ware Shoals
Waterloo
Greenwood
Cross Hill
Ninety Six
Lake Greenwood State Park
Belfast Wildlife Management Area
Newberry
Prosperity
Dreher Island State Park
Lake Murray
Gilbert
Lexington
Red Bank
Oak Grove
Columbia Metropolitan Airport
Cayce
Columbia
Pine Ridge
Gaston
Swansea
Hopkins
St. Matthews
Cameron
Elloree
Santee State Park
Santee
Summerton
Lake Marion
Santee National Wildlife Refuge
Eutawville
Cross
Pineville

Hatchery Wildlife Management Area
Canal Wildlife Management Area
St. Stephen
Moncks Corner
Francis Marion National Forest
Awendaw
Buck Hall Recreation Area
McClellanville
Bulls Bay
Cape Island

Appendix L

**Total Eclipse 2024:
A First Look**

What just happened? It's August 22, 2017. Citizens of and visitors to the U.S. are abuzz about yesterday's celestial event. Videos, photos, and social media reports abound, and traffic still isn't back to normal yet. But you missed it. Why doesn't matter. Whether you were serving on a submarine at the bottom of the Pacific Ocean, just awoke in a hospital following a zombie apocalypse, or were hampered by the thickest (and darkest!) clouds anyone ever saw, you failed to experience the awesome wonder of the 2017 total solar eclipse.

Now what? Well, you can check Appendix C to see when Earth will again experience totality. The list goes through 2030. Perhaps you'll want to plan a trip to South America or Antarctica. I urge you, however, to look carefully at the entry marked April 8, 2024. That's the next total solar eclipse that crosses the United States. And although six years and seven months (plus 19 days) sounds like a long time, it's much shorter than the average time between most eclipses occurring in a specific location. Solar eclipses only occur when the Sun and the Moon lie at the same node.

Following are the details of the 2024 total solar eclipse. And because it's likely that you won't read this again for, oh, six and a half years, I've done a short lead-in that contains eclipse basics.

The Basics

I like to think of total eclipses as examples of sublime celestial geometry. Each one is an exact lineup of the Sun, the Moon, and Earth (for a total solar eclipse) or the Sun, Earth, and the Moon (for a total lunar eclipse). And although total solar

© Springer International Publishing Switzerland 2016
M.E. Bakich, *Your Guide to the 2017 Total Solar Eclipse*, The Patrick Moore Practical Astronomy Series, DOI 10.1007/978-3-319-27632-8

L Total Eclipse 2024:A First Look

eclipses occur more often than total lunar ones, more people—actually, pretty much everyone—has seen a total eclipse of the Moon. Few, on the other hand, have seen a total solar eclipse.

The reason is quite simple: We live on Earth, and it's our perspective that interacts with the geometry of these events. During a lunar eclipse, anyone on the night side of our planet under a clear sky can see the Moon passing through Earth's dark inner shadow. That shadow, even as far away as the Moon, is quite a bit larger than the Moon, so it takes our satellite some time to pass through it. In fact, if the Moon passes through the center of Earth's shadow, the total part of the eclipse can last as long as 106 minutes. Usually totality doesn't reach that duration because the Moon passes either slightly above or below the center of the shadow our planet casts.

Conversely, the Moon and its shadow at the distance of Earth are much smaller; so small, in fact, that the shadow barely reaches our planet's surface. Anybody in the lighter outer region of the shadow (which astronomers call the penumbra) will see a partial solar eclipse.

The lucky individuals under the dark inner shadow (the umbra) will experience—a much better word than "see"—a total solar eclipse. Sometimes, only the Moon's penumbra falls on Earth, and the eclipse is partial everywhere. Not in April 2024.

A question people often ask is, "Isn't the Sun a lot bigger than the Moon, so how does the Moon cover it so exactly?" Yes, the Sun's diameter is approximately 400 times larger than that of the Moon. What a coincidence that it also lies roughly 400 times farther away. This means both disks appear to be the same size.

Regarding timing, all solar eclipses happen at New Moon. Unless the Moon lies between the Sun and Earth, it can't block any of our star's light. The only lunar phase when that happens is New Moon.

But why doesn't a solar eclipse happen at every New Moon? The reason is that the Moon's orbit tilts 5° to the plane formed by Earth's orbit around the Sun, which astronomers call the ecliptic (because that's the only place eclipses can occur).

Most of the time, our satellite is either north or south of the ecliptic. But during each lunar month, the Moon's orbit crosses that imaginary plane twice. Astronomers call these intersections nodes.

Solar eclipses only occur when the Sun and the Moon lie at the same node. Unfortunately, during most lunar months, the New Moon lies either above or below one of the nodes when the Sun is there, and no eclipse happens. On average, a total solar eclipse occurs somewhere on Earth about once every 16 months. But the average length of time between two total solar eclipses at a specific location on Earth is much longer: 330 years in the Northern Hemisphere and 550 years for locations south of the equator.

The difference between the hemispheres is due to two factors: (1) More eclipses occur during summer months (more hours of daylight); and (2) the Northern Hemisphere lies farther from the Sun during its summer, making our daytime star a smaller target (hence, easier to cover). The maximum length of totality also varies from one eclipse to the next. The reason comes from the fact that Earth is not always at the same distance from the Sun and the Moon is not always the same distance from Earth. The Earth-Sun distance varies by 3 percent and the Moon-Earth distance by 12 percent.

The result is that the Moon's apparent diameter can range from 7 percent larger to 10 percent smaller than the Sun. A bigger apparent size for the Moon and a smaller one for the Sun equals a longer totality. But a Moon that looks smaller and a Sun that appears larger means that you'll experience a shorter time in the dark.

According to Belgian astronomer Jean Meeus, the maximum duration of totality from 2000 B.C. to A.D. 3000 is 7 minutes and 29 seconds. That eclipse will occur July 16, 2186, so don't get too anxious.

The maximum length of totality during the April 8, 2024, eclipse—4 minutes and 28 seconds—won't be that long, but it's still a worthy chunk of time.

Everyone in the contiguous U.S. will see at least a partial eclipse. In fact, if you have clear skies on eclipse day, the Moon will cover at least 16.15 percent of the Sun's brilliant surface. And that's from Tatoosh Island, a tiny speck of land west of Neah Bay, Washington. But although our satellite covering part of the Sun's disk sounds cool, you need to aim higher.

Likening a partial eclipse to a total eclipse is like comparing almost dying to dying. If you are outside during a solar eclipse with 16 percent coverage, you won't even notice it getting dark. And it doesn't matter whether the partial eclipse above your location is 16, 56, or 96 percent. Only totality reveals the true celestial spectacles: the two diamond rings, the Sun's glorious corona, 360° of sunset colors, and stars in the daytime. But remember, to see any of this, you must be in the path.

That said, you want to be close to the center line of totality. The fact that the Moon's shadow is round probably isn't a revelation. If it were square, it wouldn't matter where you viewed totality. People across its width would experience the same duration of darkness. The shadow is round, however, so the longest eclipse occurs at its center line because that's where you'll experience the lunar shadow's full width.

And, just like for the 2017 eclipse and every other predicted eclipse, I want to stress that this event will happen! As astronomers, some of the problems we deal with are due to the uncertainty and limited visibility of some celestial events. Comets may appear bright if their compositions are just so. Meteor showers might reach storm levels if we pass through a thick part of the stream. A supernova as bright as a whole galaxy maybe visible, but you need a telescope to view it. In contrast to such events, this solar eclipse will occur at the exact time astronomers predict, along a precisely plotted path, and for the lengths of time given. Guaranteed. Oh, and it's a daytime event to boot.

2024 Specifics

The Moon's shadow first strikes Earth in the Pacific Ocean just north of Penrhyn Island, one of the Cook Islands. That location will experience a 98-percent partial eclipse. Seventy-three minutes later, totality first strikes land at Socorro Island, part of the Revillagigedo Islands, which lie roughly 370 miles (600 kilometers) west of Mexico. If, for some reason, you choose that location to view the eclipse, be sure to position yourself at the island's far southeastern tip—you'll enjoy an extra 34 seconds of totality (the span is 3 minutes and 36 seconds) there.

The shadow's path covers a few more tiny islands before it encounters North America just southeast of Mazatlán, Mexico. Viewers from that location will enjoy 4 minutes and 27 seconds of totality. And if you wish to stay in Mazatlán, you'll lose only 10 seconds off that span. Greatest duration, 4 minutes and 28.1 seconds, occurs when the shadow reaches San Martin, roughly half the way from the coast to the Mexican border with Texas. In fact, the duration of totality is never more than 1 second less than this maximum during the shadow's more than 550-mile (885 kilometers) voyage through Mexico.

Totality first occurs in the U.S. as the shadow crosses the Rio Grande River at the wonderfully named Radar Base, Texas, which lies in Maverick County. There, the duration of totality equals 4 minutes and 27 seconds. As the eclipse progresses in the Lone Star State, a huge number of people won't have to travel anywhere to see it. That said, just a few miles journey to the center line can increase their durations of totality. San Antonio, Austin, Waco (Baylor University will enjoy 4 minutes and 10 seconds of darkness), Dallas, and Fort Worth all lie under the shadow, although none is on the center line. Still, that's more than 11 million people who can experience the eclipse with little to no effort. And we're not even out of Texas yet.

The center line then passes through Oklahoma, Arkansas, Missouri, Illinois, Indiana, Ohio, New York, Vermont, and Maine. Those wishing to observe the eclipse from the same location the center line crossed during the August 21, 2017 eclipse should head to a location near Makanda, Illinois, which lies just south of Carbondale. A word of warning, if I may: The weather in Illinois in April—and here I'm specifically talking about cloud cover—is a far cry from what it is in August. Your chances of actually seeing the eclipse increase dramatically as you head to the southwest. Not to mention that you'll pick up an extra 15 seconds of totality from center line locations near San Antonio.

Cities in the path include Little Rock, Arkansas; Indianapolis, Indiana; Dayton and Cleveland, Ohio (with northwestern parts of Cincinnati and Columbus under the shadow); Buffalo, Rochester, and Syracuse, New York; and about half of Montreal, Québec, Canada.

U.S. Eclipses Beyond 2024

For those wanting to view a total solar eclipse in the contiguous U.S. after the 2024 event, it's a 20-year wait until August 23, 2044 for the next one. That eclipse is visible only in northeastern Montana and a tiny segment of North Dakota. Its greatest duration of totality, 2 minutes and 4 seconds, happens over Canada's Northwest Territories.

At the intersection of the center line and the U.S.–Canada border, totality is 20 seconds shorter: 1 minute and 44 seconds. The small towns near that point, namely Hogeland and Turner, Montana, experience totalities just a few tenths of seconds less. And Chinook, Zurich, Harlem, Dodson, and Malta lose only another second.

But here's the potential problem in 2044: The Sun's altitude at the border is only 4.7° at mid-eclipse. And at Malta, it's a paltry 3.6°. Even if 99 percent of the sky is clear, low-lying clouds, haze, or smoke from forest fires could hamper your viewing.

Three more total solar eclipses track through the contiguous U.S. in the 21st century. And if 2044's eclipse disappointed you, you'll be thrilled by the one that happens less than 1 year later. The event August 12, 2045 is a truly spectacular cross-country eclipse with totalities lasting 4 minutes and 23 seconds for the northern California coast to its amazing maximum of 6 minutes and 6 seconds at Port St. Lucie, Florida.

The center line of the total solar eclipse March 30, 2052, only lands on Florida and Georgia, but totality in that small path in the U.S. lasts between 3 minutes and 30 seconds in Savannah, Georgia, and 3 minutes and 44 seconds a bit northwest of Laguna Beach, Florida.

The final total solar eclipse whose path intersects the contiguous U.S. occurs May 11, 2078. Like the one in 2052, it also tracks through the southeastern U.S. They'll be partying in New Orleans for this one (like people there need another reason!) because totality over the Big Easy lasts a whopping 5½ minutes! From there, the Moon's shadow tracks northeastward across Mississippi, Georgia, and South Carolina, finally heading off land into the Atlantic Ocean after producing a totality of 5 minutes and 17 seconds for those in Nags Head, North Carolina.

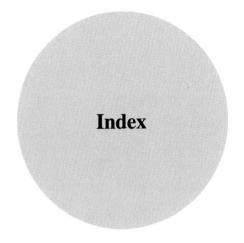

Index

A
Altitude, 7, 8, 26, 39, 62, 63, 79, 90, 95–97,
    140, 141, 261, 288–292, 294–308
Anderson, Jay, 31, 259, 260, 262
Annular, 40, 41, 47, 50, 52, 59, 60, 75,
    107, 237
Anomalistic period, 38, 39
Aphakia, 78
Aphelion, 8, 9
Apogee, 8
Azimuth, 8, 9, 140–142

B
Baily's beads, 9, 60, 102, 103
Barlow lens, 153–155, 172, 174, 175
Binoculars, 22, 76, 95–97, 107, 113–128,
    140, 153, 155, 167, 170, 205–211,
    243, 244
    mounts, 123–125

C
Catadioptric, 136–139, 143, 167
Celestron, 115, 119, 122, 123, 125, 130, 131,
    133–142, 144–149, 152–154
Center line, 10, 20, 26, 33, 52, 56, 59,
    223–228, 239, 254, 259–286, 291,
    292, 295, 296, 299, 300
Chou, Ralph B., 73, 78, 79

Chromosphere, 10, 12, 104, 170, 172, 174,
    176–179
Collimate, 135
Corona, 5, 10, 11, 22, 43, 49, 51, 52, 58,
    62–64, 67, 68, 70–72, 95, 99, 100,
    102, 103, 108, 127, 183, 186, 189,
    190, 193, 194, 239, 244

D
Daystar (filters), 170–175
Declination, 39, 90, 91
Diamond ring, 5, 11, 103, 108
Digital single-lens reflex camera (DSLR),
    157–160, 181, 187, 191
Draconitic period, 37

E
Ecliptic, 11, 13, 37
Eddington, Arthur, 63, 65
Einstein, Albert, 63, 65
Equinox, 39, 54
Exeligmos, 39
Eyepiece(s), 77, 82, 85–87, 115, 117, 118,
    120, 122, 129, 133, 137, 143, 147,
    149, 150, 152–156, 166, 168, 171,
    172, 175–177, 179, 180, 189, 201,
    202, 206
Eye relief, 118–119

© Springer International Publishing Switzerland 2016
M.E. Bakich, *Your Guide to the 2017 Total Solar Eclipse*, The Patrick Moore
Practical Astronomy Series, DOI 10.1007/978-3-319-27632-8

Made in the USA
Columbia, SC
11 July 2017